Environmental Issues

Environmental Issues

Behavioural Insights

EDITED BY

PARUL RISHI

RAWAT PUBLICATIONS

Jaipur • New Delhi • Bangalore • Mumbai • Hyderabad • Guwahati

ISBN 81-316-0261-3

Published by
Prem Rawat for **Rawat Publications**
Satyam Apts, Sector 3, Jawahar Nagar, Jaipur 302 004 (India)
Phone: 0141 265 1748 / 7006 Fax: 0141 265 1748
E-mail: info@rawatbooks.com
Website: www.rawatbooks.com

New Delhi Office
4858/24, Ansari Road, Daryaganj, New Delhi 110 002
Phone: 011 2326 3290

Also at Bangalore, Mumbai, Hyderabad and Guwahati

Printed at Chaman Enterprises, New Delhi

Contents

Preface

Environment and development are among the foremost concerns of today's globalized world. At one end we are eager to achieve the materialistic targets by creating newer and advanced luxuries of transport, communications and comforts, at the other, we have started experiencing the effects of these developments on global climate and environment. So, it is a real trade off that we have to decide and human beings play a major role in this regard.

This book is a constellation of various thematic perspectives linking human behaviour to environment for global well-being. It is divided into five sections, each dealing with a different dimension of human behaviour as linked to environment.

Part I on 'Environment and Development' starts with the musings of a dinosaur on some fundamental issues addressed by an environmental scientist. It is followed by the psychological analysis of limiting resources and consumptive behaviour. Next chapter in this section deals with behavioural perspectives of some developmental issues of urbanization and how they lead to climate change. It is followed by urban environmental stress and how to cope with that for good quality of life and well-being.

The major turning point to tackle these human-induced environmental problems is the mindset. The Part II on 'Environmental Attitudes and Behaviour', addresses the nature and formation of environmental attitudes and how to change the mindset which is not pro-environmental. It also talks about the linkage between attitudes and behaviour.

From micro-level environmental attitudes, we proceed to macro level concerns of the business world in terms of environment. The Part III on 'Ethics, Business and Environment' is the summation of conceptual models of corporate social responsibility and environmental ethics with the case of environmental initiatives and practices of the Hind Lamps, a Bajaj group organization.

Macro-level corporate initiatives are the result of effective leadership practised in the business world. The Part IV on 'Environmental Leadership' combines the Bolivia case of challenges of environmental policy and leadership with the practical skill based approach to environmental leadership based on Smith Sonian Environmental Leadership course.

The Part V culminates the above and says that spiritual models and Indian cultural practices help promote positive environmental appraisal and mindset for global well-being. This approach is substantiated with the help of a case of Dayalbagh, a spiritual institution practising better worldliness.

Finally, there are comments on some key dimensions of human behaviour which have a strong bearing on humanizing the planet for global well-being.

Acknowledgements

First of all, I bow my head in front of the Almighty who made me capable of taking this work, provided persistence to continue with it in spite of all odds and gave me the wisdom to do justice with it.

I gratefully acknowledge all the authors, who have contributed to enrich this book. Michael Philip, Leslie Wildesen, Juan Pablo, Ananda Das Gupta, Nachiketa Tripathi, Parag Dubey, Swati Moghe, Jeanne C. Crough and Gayatri Khuntia. All the authors have given their best academic and experiential inputs for this book.

I extend my heartfelt gratitude to Dr D.K. Bandyopadhyay, Director, IIFM, for motivating and supporting me to complete this volume. Gratitude is also due to Dr Suprava Patnaik for encouraging me to take up this work and extending utmost support during initial phases.

I also acknowledge the support provided by Mr T.K. Patro in the matter of typesetting and formatting this book. Graphics' designing and final formatting by Mr K.V. Chandrasekhar is also sincerely acknowledged.

I also express my special gratefulness to Prof (Ms) S.P. Sinha, my teacher, PhD guide and mentor for her unconditional academic and personal help in every sphere of my life.

The most important partner in writing and editing this book has been my husband, Dr Rishi Nigam. He was always with me extending unstinting support and encouragement which made it possible to complete this book. My loving daughters, Tarana and Siddhi, also deserve a special note of appreciation. Their smiling faces were instrumental in re-energizing me for the endeavour.

Finally, I owe a deep gratitude to my parents and family members who were strongly instrumental in my each and every success and accomplishment.

Parul Rishi

Contributors' Profile

Ananda Das Gupta is active in the areas of ethics in business, corporate social responsibility, organizational development and strategic human resource management by way of research, teaching and consultancy/training for the last twenty years. He is involved with the Social Responsibility Research Network of England, being led by Prof David Crowther. Recently, he edited *International Journal of Social Economics* of Emerald Group, England as a Guest Editor for the Special Issue on India. Besides having chaired international seminars on CSR Issues in India, London and Kualalumpur, his book *Human Values in Management* was published from Ashgate Publishing England, and his other book: *Corporate Citizenship: Perspectives in the New Century* has very recently come out from Cambridge Scholars Publishing, England. Dr Das Gupta is now invited to be another guest editor for a special issue of *Social Responsibility Journal*, England. He is Associate Professor (HRD-Area) at Indian Institute of Plantation Management in Bangalore. Email. adg-iipm@vsnl.net

Gayatri Khuntia is M.Sc. Environmental Sciences, M.Phil. (Natural Resource Management) from Indian Institute of Forest Management and consultant, environment and rural development and gender issues. Email. gayatri_khuntia@yahoo.co.in

Jeanne C. Crouch, M.S., is a co-founder and Principal of ETCI, with decades of experience in international banking, business development, and leadership management. Her expertise can be seen in all aspects of ETCI's successful training and consulting models, and long-term client relationships. She is the co-designer and a popular trainer of the Smithsonian Environmental Leadership Course. Email. jcrouch@bealeinternational.com

Juan Pablo, M.Sc., is Director of the Latin America and Caribbean Department of NatureServe in USA, with over 20 years of experience working in conservation of natural resources with a strong background in conservation biology, environmental policy, biodiversity conservation, and biodiversity information management focused primarily on Latin America's sustainable development and local capacity building initiatives. Former Vice-Minister of Natural Resources and Environment under the Sustainable Development and Environmental Ministry of Bolivia, and Chairman of the National Trust Fund for Environment (FONAMA) in Bolivia. Former country programme director of Conservation International in Bolivia. Undergraduate degree studies in Biology from the University of San Andres in Bolivia and Master Degree in Rural and Land Ecology at ITC in The Netherlands, and postgraduate studies in Global Studies at the University of Denver. Email. juanpablo_arce@natureserve.org

Leslie E. Wildesen, PhD Wildsen is a co-founder and the Principal of ETCI, and an internationally recognized expert in resource management policy and programme development, environmental leadership, environmental impact assessment (EIA) and environmental management systems (EMS). Since 1967, she has participated in every aspect of environmental impact assessment from initial resource inventory to EIS preparation and policy development, and has published over 150 papers. Her decades of experience include service in both Federal and state agencies, and she is a staff member of the US House of Representatives. She has served as an appointee of the President of the United States, the Governors of Oregon and Colorado, and two secretaryships of the Interior on various technical and advisory panels. She has also served as an officer or board member of numerous professional associations, including the National Association of Environmental Professionals, where she chaired the NEPA training committee from 1996–1999. Email. lew@envirotrain.com

Michael Stuart Philip, O.B.E., M.A. (Oxon), B.Sc. (London) served in Uganda as a member of the Colonial Forest Service, in charge of silvicultural, utilization and forest protection research. Joined the

Department of Forestry in the University of Aberdeen; retired as Reader in Forestry. Seconded to the Norwegian Development Aid Agency at University of Dar-es-Salaam. Appointed by the DFID as Head of the Nepal-UK Forest Research Project. Visited DFID's rural development project in Thailand. Worked as an international forestry consultant in India, Nepal, Sri Lanka, Malawi and Brazil. Editor, *Forests Trees and Livelihoods* and Author of *Measuring Trees and Forests,* CAB International, Oxford 1994. Co-author with H.C. Dawkins of *Tropical Moist Forest Silviculture and Management: A History of Success and Failure.* CAB International, Oxford. 1998. Email. philipfor@aboyne93.fsnet.co.uk

Nachiketa Tripathi, did his Masters from Kanpur University and obtained PhD in Social/Environmental Psychology from IIT Kanpur in 1996. Thereafter, he joined Centre for Organization Development (COD) Hyderabad. Later, he moved to IPM, Kanpur. Dr Tripathi joined IIT Guwahati in 1998 and presently he is an Associate Professor, teaching and doing research in the areas of environmental psychology, organizational behaviour and human resource management. He has scholarly papers in reputed journals and made presentation at several national and international conferences, such as APA Conventions, ICAP, ICP and ECP. Email: nachi@iitg.ernet.in

Parag Dubey is an Assistant Professor in the Faculty of Marketing Management at Indian Institute of Forest Management, Bhopal. He has been teaching courses on marketing, marketing research, green marketing, rural marketing and marketing of forest products to postgraduates, and providing training on Strategies of forest products marketing to working professionals. His areas of research interest include marketing of forest products particularly, non-timber forest products, marketing of environment friendly products, and corporate social responsibility. He has been publishing on these issues nationally and internationally. Had been awarded Japanese government scholarship for research project on Green Consumerism at Kobe University, Japan in 2002. Email. parag@iifm.ac.in

Parul Rishi, PhD in Behavioural Science and is teaching in the Faculty of HRM at Indian Institute of Forest Management, (IIFM), Bhopal for about one and a half decade. Witkins Okonji Fellow, 2008, IAUH, Canada Fellow, 2008, CIDA Fellow, 2007, ITTO, Japan Fellow of 2004 for participating in Smithsonian Environmental Leadership Course at USA, and ARTS Fellow on Environmental Psychology by International Association of Applied Psychology, at Xi 'an, China.

Recipient of Young Scientist Award of Indian Science Congress, ICSSR doctoral fellowship and national scholarship of Ministry of HRD. Engaged in teaching, training, research and consulting assignments focusing primarily on behavioural perspectives of forest and environmental issues of stress management, environmental and forestry leadership and corporate social/environmental responsibility. Published around 60 papers in international/national journals and book chapters, and has presented papers in different scientific national and international conferences. Email. parul@iifm.ac.in

Swati Moghe is M.Sc. Forestry, and prospective PhD (Forest Research Institute). Presently working as a freelance consultant for projects of forestry, environment and social sciences with a special focus on quantitative analysis. Worked with Indian Institute of Forest Management as Senior Research Fellow for various projects and consultancies related with social and behavioural issues of forest and also worked with ORG-MARG, Nepal, as consultant for World Health Project. Email. swatimoghe@yahoo.com

Part I

Environment and Development

EDITOR'S INTRODUCTION

Environmental issues have been areas of concern since many decades. Yet, somehow, we are not able to resolve some of the major issues, notwithstanding high level research, media attention, environmental campaigns and global environmental meets. This scenario forces us to think where the knot of the problem lies. Are we seriously concerned about the environment and to resolve some of the major prospective environmental crises like global warming and climate change? Or it is just a 'business as usual' approach for us? Our level of seriousness depends on many issues starting from who is responsible? What I can individually do or in a group? In what way will it impact the global environment? After how long the negative effects of consumerist behaviour will be visible? And so on.

As all the environmental issues are deeply linked with development, the Part I on environment and development addresses some of the basic issues linking the two. Urbanization, industrialization, globalization and materialism are closely associated with the shift from a closed, subsistence economy to an open, market-oriented economy. The forces associated with mass production, mass marketing and advertising, great spending power, ease of communication and modern technology, somehow or other threaten the very basis of sustainability of environment. Day by day increasing materialism and consumptive behaviour are limiting the available natural resources, thereby, are endangering the very survival of common pool of resources. The following lines on globalization are worth to be quoted here.

These lines remind us of our ethical responsibility towards nature and environment. One of the major challenges that global environment has started facing, is changing climate. Human environment interactions involve not only the question of resource use per person, but also our ability to understand the science of the environment, our ability to regulate our impact on the climate, our beliefs and values, our attitude to the future, particularly risk, and our ability to negotiate solutions at both the local and global levels regarding climatic change.

Moreover, there are many issues besides global warming, which are associated with development and may lead to urban environmental stress. The core dilemma of urbanization has its impact not only on environment but also on the physical and mental health of human beings. The urbanized societies have to bear a lot of stressors at the cost of peace and harmony, for which efficient coping on the part of people is recommended to buffer the negative outcomes of stress.

The four chapters of Part I conceptually link environment and development related issues with a focus on how behavioural science can play a role in managing the depletion of resources, consumerist behaviour, climatic change and above all the urban environmental stress, for global well-being.

1

People and the Environment

Musings of a Dinosaur

MICHAEL S. PHILIP

So much has been written about the problems of the environment and so many scientists and 'experts' are involved—all with their own particular backgrounds, viewpoints, goals and benefits—that it is difficult to obtain an objective, holistic account. The complete scene and the links between the main elements of mankind and his environment—the physical, climate, ecology, human populations, governments, civil organizations, cultures, trade, materialism, technology, urbanization etc., are difficult to appreciate. Breaking up the whole system and considering the parts separately is misleading as it distorts the picture and obscures the relationships and weights of the different parts. Even though the whole is complex and little understood, fragmentation—by geographical regions and sub-problems-results in oversimplification and a loss of balance.

In many parts of the world and in many societies awareness of the problems of climatic change, environmental degradation, pollution, loss of biodiversity etc., has grown enormously until now, at least in the UK, one cannot open a paper, hear the news, listen to a sermon etc., without finding a reference to one or more of these problems and perhaps this is equally true elsewhere. However, many of the claims that action is being taken and 'degradation' is being halted or reversed, may be very misleading because they are not considered in either a global or holistic context.

Reading some of the literature reminds us of the warning—that when consulting articles and, especially, sites on the worldwide web—always to bear in mind the questions:

- Who wrote it?
- Why was it written?
- Under what circumstances and pressures was it written?
- Who will gain from it being read?

Outlook of an author is always coloured by his own memories, perceived 'truths', experience and, perhaps, a touch of his personality which in this case is pessimistic about the human race.

Land Use and Natural Resources Policies: An example of a past failure to see the obvious.

After the 1939–45 war, the effect of the increase in population on the availability of resources, land use allocations and forest management policies was not mentioned. Land use policies—especially in the countries dominated by colonialism—remained much as they had been set in 17th century Europe. These policies were transferred by the colonial powers to many tropical countries (Fay and Michon, 2005). Even up to the 1960s, forest managers, in many developing countries, though aware of the projections of an increasing demand for wood products based, in part, on forecasts of the growth in the population, still failed to question the concept of a 'permanent forest estate' and its implications of the centralized control and a stable, if not static, demand for land for growing food. This was in spite of the population's reliance on subsistence farming and exponential rate of increase. (With hindsight, this seems incredible but, perhaps, is no less incredible than our failure to forestall global warming.) Not surprisingly, this rigid land allocation policy failed the test of time and something more flexible and decentralized is only now slowly evolving.

The declaration of a 'permanent forest estate' allowed the state to assume ownership of the forests that were declared to be of national importance—either for the protection of the soil and water supplies, or for their wood and revenue production. People who traditionally had used the forest areas for their needs for fuel, building materials, grazing, collecting fruits and other 'minor forest produce' were granted rights or privileges that, generally, permitted them to continue their customary harvesting practices to support their livelihoods, but clearing of forest for agriculture and commercial dealing in other forest products were prohibited. As population increased, pressure on the forest increased to such an extent that the forest services became alienated from the people, forest laws were ignored and sustained management became impossible. Pressure from local people for local goods and services conflicted with the trend towards timber harvesting by large commercial concerns that were backed by politically influential entrepreneurs and financiers—who claimed to stimulate development, raised employment, pay wages, improve infrastructure in transport, communications, health, education etc., and earn foreign exchange through exports of logs. The benefits from this industrialization were expected to flow down to the poor and ameliorate the lot of the poor (Westoby, 1962). Facilities for the manufacture of sawn timber, panel products and finished goods were introduced to raise forest product values. Industrialization was supported by the influences of urbanization that was associated with an expansion in the global economy and increased use of manufactured wood products; eventually this resulted in home markets for timber and panel products replacing exports in many countries. In other instances, countries that earned revenue from timber-based exports are now importing manufactured wood products.

In recent years there has been a gradual loosening of the reins of central governments in many developing countries, accompanied by a strengthening of local authorities and empowerment of communities to manage their natural resources. This decentralization has been accompanied by revision of land tenure and the recognition of traditional individual and communal tenurial systems (Wiley, 2004). Unfortunately, locally, the benefits from resources are sometimes garnered by elites rather than shared equably through the local community. The conflict of interests between the needs of those on a subsistence economy, those demanding manufactured goods and those demanding the services of the forest on an international scale has yet to be resolved at local, national or international levels, and can only be tackled through continuous negotiation as needs, values and bargaining powers change

With hindsight, the concepts behind the policy of reserving a 'permanent forest estate' were obviously unsuited to the situation with an ever-increasing human population and the development of the worldwide trade in wood and, especially, ply wood and panel products. Are we going to continue to refuse to foresee the results of obvious trends and repeat this type of error in the future?

THE FUTURE

Urbanization, industrialization, globalization and materialism are closely associated with the shift from a closed, local, subsistence economy to an open, market-oriented economy. The forces associated with mass production, mass marketing and advertisement, greater spending power, economies of scale in the provision of goods and services, ease of communication and contact, the spread of information through modern technology etc., all promote industrialization, globalization and urbanization, and feed demands for material goods, their transport and advanced services in health, education and leisure activities. The price is an ever-increasing demand for water and energy and an ever-increasing degree of pollution of the environment. Globalization tends to lead to unsustainable, destructive, economic growth, decrease in governments' ability to control growth, increase in corporate power, stimulation of polluting sectors, such as transport, increase in probability of economic crises, decline in traditional controls, separation of action and responsibility from effect, ascendancy of the growth paradigm, loss of diversity like homogenization of cultures, loss of species and varieties through monoculture, narrow selection etc.

Many of the basic facts of the present situation of the world are clear, although the complexity of the system renders the combined trends, rates of change and interactions obscure, and prediction fraught with difficulty and uncertainty. Perhaps with oversimplification, I contrast the 'Gaia' and the 'Doom' theories.

The *Gaia* Theory

The exponents see the planet as self-regulating and quote the example of the earth's control of its temperature in spite of the fluctuating or increasing heat of the sun. Other parts of the system that appear to be self-regulating over long periods of time are the salinity of the seas and the carbon dioxide cycle. These cycles link living and non-living components in bio-transport systems—the carbon and water cycles for example. One corollary of the Gaia theory of self-regulation is that damage once done is very difficult to repair (Lovelock, 1979).

The Doom Theory

Simply stated, exponents of the Doom theory see *Homo sapiens* as the agent of its own destruction through the use of resources in an unsustainable manner that results in pollution of the system to the extent that human life becomes impossible. Dryzek (2005) refers to this as the pessimistic outcome of survivalism.

THE ELEMENTS IN THE PROBLEM

Some of the components of the system over which we have a degree of control are:

Homo Sapiens

What is the nature of the beast? Is it, as Dawkins (1979) infers, driven by the selfish gene? Or has mankind a higher destiny? What part will be played by the contrasting elements in human nature-good and evil; altruism and selfishness/greed; selflessness/cooperative and competitiveness; peace and violence? Can the higher elements in Maslow's (1943) triangle of needs be brought into play in time to save the planet?

Can the world's religions promote the positive sides of human nature? The debate over the role of 'faith' and 'scientific rationalism' has taken a 'Green' hue in recent times, but we see no indications in our current global society that the responsibilities delegated to humanity by a Creator are matched by altruism, charity, and sense of responsibility for the environment, or are diluting materialism, reducing envy, or affecting a significant reversal of consumption and pollution. We may doubt the efficacy of Bill Clinton and others' attack on poverty—though we may applaud their sentiments, whilst the success of the *Grameen Bank*, Bangladesh initiated by Mr Muhammad Yunus—recently awarded the Nobel Peace Prize shows that the positive sides of human nature can be promoted within capitalism.

Population[1]

The world population has risen from some 2 billion in 1925, to 3.5 billion in 1965 and is now approaching 7 billion and is predicted to reach nearly 9 billion by 2050. While the increase in the population of the industrial countries of the West is small, the major part of the increase is predicted to occur in Asia—with India likely to overtake China within 30 years and become the country with the highest population in the world. However, these predictions are fraught with uncertainty and may present a distorted picture of the future. The effects of changes in education, health services, poverty alleviation etc., are equally unclear.

Differing birth and mortality rates of nations affect their demographic structure—with the industrialized West tending to have an ageing population exacerbated by falling birth rates and increasing longevity, contrasting with less developed countries with a high birth rate and falling infant mortality; decreasing life expectancy through HIV/AIDS is leading to a lowering of the rate of population increase in some lands. So, not only are the numbers changing but so are the structures of the population. As the young and the old have very different time horizons, aspirations etc., they assess the future from completely different viewpoints.

The human species is unique in the history of the world on account of its numerical dominance, flexibility, technology and ability to exploit, manipulate and alter the environment for its own use. Will the species bear out the predictions of the Gaia Model—that natural selection assists those organisms that are in balance with their surroundings—or those of the Doom Theory?

Though the consumption of resources and contribution or pollution per head of many developing countries is low, high populations raise their total contribution to the problems of the environment to a level comparable with that of countries with fewer people but higher levels per head; at the same time their potential for increased rates of consumption and pollution associated with the development of their economies, is high. Can their development be sustainable?

Carbon Emissions and Global Warming[2]

The recent findings of temperature and carbon dioxide concentrations in the Vostok Antarctic ice cores establish the correlation (but *not* causation) between the two, and the level of fluctuations that have taken place over the last 400 millennia. The increase in carbon dioxide in the atmosphere is part of the pollution resulting in the greenhouse effect that is associated with rising temperatures.

Concentrations of carbon dioxide are now at the highest level in this period, having risen from some 250 ppm prior to the industrial revolution (c 1850) to some 375 ppm now. Similar high concentrations have occurred before—cyclically at around 1,00,000 year intervals, *but not accompanied by the current extremely high rate of increase.* The extent to which the current increase can be attributed to mankind is uncertain—perhaps currently around 10 per cent and rising, but 50 per cent of the man-made emissions originate in industry and transport. The levels of carbon dioxide fossil fuel related emissions have risen from 100 million tons in 1900 to around 25,000 million tons in 1995; emissions rose by a further 25 per cent between 1994 and 2004. The US is responsible for 20 per cent, China 10 per cent and the shares of Russia, India and Brazil are rising rapidly. Annual rates of over 20 tonnes per head in US, 10 in UK, 2 in China, 1 in India and less than 0.5 in subsistence economies underline the contribution from industrialization.

Many countries with large populations, such as India, China, Brazil and Mexico etc., that currently have a low level of emissions per head of population, will—I foresee—increase their total emissions as the effects of their urbanization, industrialization and materialism take hold of an increasing proportion of their population.

In recent months, governments, backed by the newspapers and other sources of information, have appealed to domestic users to cut their consumption and wastage of energy by such simple means as using the energy saving light bulbs, turning off the standby modes of many domestic appliances etc. Political parties have climbed on the 'Green' wagon and advocated restricting carbon emissions from power plants and industry and switching to renewable sources of energy, wind and wave power. The trade in credits for carbon emissions is developing. Meanwhile, the debate on the long-term danger from the by-products of energy from nuclear plants and how to store them safely continues. Manufacturers have given greater attention to alternatives to petrol and diesel—hydrogen and bio-ethylene, for example for transport. However, there is little information on the cost in energy and pollution of the manufacture of the equipment and processes to supply the alternatives, nor on the effects of such developments relative to the demands engendered by the forces driving up consumption from the developing world.

The degree to which variation in the emissions of the sun are responsible for the rise in the earth's temperature is unclear, though it is likely that such variations were responsible for equivalent variations in the temperature of the earth associated with the various Ice Ages and intermittent warm periods. Nevertheless, to my mind, the crucial fact is the current high rate of increase in levels of carbon dioxide in the atmosphere—due to, or at least associated with, transport and industrial emissions. I see little likelihood of this trend being reversed in the next twenty years.

A separate issue is the damage to the ozone layer over the Antarctic and Arctic regions due to the release of Chloro-Fluoro-Carbons (CFCs) and nitrogen oxides. The holes lead to rising ultraviolet radiation reaching the earth's surface. Most of the chlorine and 50 per cent of the bromine in the stratosphere has been released by human activities. Although the production of the main CFCs is now banned under the Montreal Agreement, the extent of the problem is not, or is only slowly, diminishing and active research deserves a high priority.

Global warming reflects in one measure some of the changes that the human species is precipitating. The resultant melting of the ice fields in Arctic and Antarctic regions is changing the currents in the oceans, affecting climates and leading to an inevitable rise in sea levels; some models currently in use predict an increase in the frequency of extreme climatic events. Rising seas levels will be disastrous for low-lying areas—especially densely populated deltas and islands with little high ground. I find the loss of the snows of Kilimanjaro in Tanzania and of the glaciers on the Ruwenzori in Uganda particularly horrifying. Likewise the loss of snow cover in the Sierra Nevada must be equally frightening to the Californians.

Disasters[3]

The evidence that the world has suffered a multitude of cataclysmic events in the past—massive volcanic outpourings, glaciations, impacts of meteorites, tectonic movements etc., suggests that the global system can—in geological time—regain its balance. What is less certain is—can the human species survive in the same time scale? As yet, evidence for change in the frequency of extreme climatic events is lacking, but disturbance to the oceans' systems caused by rising temperatures and ice melt has the potential to produce both such cataclysmic events and less spectacular but no less devastating flooding along coastal fringes.

Poverty,[4] Materialism and Consumption[5]

More than half of the world's people live below the international definition of US$2 *per day* income. The industrial revolution released mankind from restrictions derived from the limits of its own and its domesticated animals' physical power through the use of machinery. The cost of this has been levied through the extent of the energy needed and the pollution caused.

So-called developed mankind has invested in development, research and services that have facilitated provision of improved health, opportunities for education etc. Less welcome outcomes have been the bloom of materialism, extreme forms of egoism, selfishness, nationalism, politics, globalization, destruction and unsustainable resource use. These now affect all of mankind, albeit to varying degrees. More beneficial developments in culture, altruism, charity etc., have done little to change the impact or the distribution of the benefits and costs of development. Poverty and inequity are universal companions of materialism.

The current rapid rise in the industrial capacity of countries like China, India, Brazil and Mexico with large populations and the potential for similar trends in less developed countries like Indonesia, South Africa etc.,—is highly likely to be accompanied by increased demands per head of population for all resources—including water, food and energy—and increased—pollution by industry and the accompanying demand for mobility and worldly goods.

On an average, an individual in an industrialised, materialistic society is consuming more than thirty times the resources of one in a subsistence economy, and will pollute the environment to a similar relative extent. To date, advances in technology have allowed people to exploit the resources demanded, and predicted shortages foreseen in the early papers from the Club of Rome have largely failed to materialize. Nevertheless the costs in terms of energy used to supply these needs are liable to rise. In the past 20 years while the population has increased by 35 per cent, auto production has risen by 45 per cent, paper consumption by 90 per cent, and expenditure on advertising by 100 per cent. At the same time, gross total production has doubled—a rate of increase that is likely to be exceeded in the next 20 years.

Urbanization[5]

Half of the world's population now live in urban surroundings. This proportion is increasing due to the attraction of opportunities for employment, income generation and better education, health and public services than in the more remote rural areas. Urban societies are more materialistic, consume more resources and cause more pollution per head than those who live at a subsistence level. As yet, there is little sign that development can be Green, though attitudes of commerce are changing—reducing emissions, employing environmental impact assessments, using sustainably managed resources, adopting socially equitable policies of fair trade and benign technologies as wind farms etc., (Angelsen and Wunder, 2006).

Food and Water[6]

Pressure from a growing population for food implies a demand for more water and additional stress on ecological systems. This will affect the poor who are most vulnerable to water scarcity. The greatest impact is likely to arise from upgrading the water efficiency use in rain-fed agricultural systems.

In the 20th century, the world's population tripled but water use rose sixfold and falling levels in aquifers mined for agricultural and urban use were becoming widespread. More effective use of water for irrigation in agriculture is essential. The technology to desalinate sea water is available but the cost is high.

Energy[7]

In the last twenty years the price of oil has fluctuated from 13 to 79 US $ a barrel—though the range is less when the data are adjusted for inflation. In the last ten years energy consumption has nearly doubled but, now, more efficient systems for domestic and industrial applications are being introduced. Currently, most energy comes from fossil fuels but there is no doubt about the practicability of other sources—nuclear, solar, wind, wave, hydrogen etc.[8] However, in order to capture any of these, some form of technology is required and these are not engineered without themselves requiring energy—so the net results are, as yet, not always favourable.

Predictions up to 2030[9] suggest that in spite of the demand likely to rise by 60 per cent, there will be no energy shortage, and that oil, gas and coal supplies will continue to be the major sources. Cleaner technologies will lower the carbon emissions per unit of energy, but the rising demand, much of which will come from the developing countries, will cause a similar rise in emissions of carbon dioxide. Advances in technology—especially carbon capture—will deliver an impact but at the cost of rising real prices. The link between the alleviation of poverty and rising demand for energy will exacerbate the problems of pollution.

Health[10]

Good health/longevity and disease/early death represent two sides of a single coin. Both affect the demographic structure of a population and both bring problems of resource allocation and management. Good health and clean water are claimed by FAO to be a human right—but the actual situation in the world is very different. In the richer, technologically more advanced parts of the world, infant mortality is falling and life expectancy is increasing, but the impact of the HIV/AIDS virus and diseases like malaria are having the reverse effect and affecting the poorer areas of the world more severely than the richer parts.

The greater the success in improving health, the greater the problems of resource allocation and pollution reduction.

Science and Technology

Man's ability to invent and use new knowledge has been the driving force of change over the last few millennia It is probable, that given time, mankind can harness as yet unused sources of energy and other resources to supply the needs of the expanding population. Again the question is: Will major disturbances to the global system occur before mankind learns to use resources in a sustainable way?

The Problem with Humanity

Why does mankind despoil the planet? What parts are played by selfishness, materialism, mimicry, greed, need to dominate, altruism, community spirit? Comparative studies of man and chimpanzees suggest that the mechanism of evolution involves a struggle for dominance, both within and between groups. The resulting conflicts provide opportunities for individuals and groups to 'feather their own nests' at the expense of others. When such activities become excessive, society acts and tries to restore the balance. Philosophers and others strive to discover, preach and practise different paths for man's progress, while the common 'man' simplistically categorizes actions as GOOD or EVIL—depending on 'his' cultural, educational and personal background. The world's religions recognize these faults in mankind and strive to inculcate a more altruistic approach to living; others rely on rationality to guide their actions.

We may not be a philosopher or seer and may be unable to answer the question, but the relative balance between the forces that cause exploitation of the planet's resources and those that promote conservation will, probably determine the future of the species.

IS THERE A WAY FORWARD?

Who are the participants?

Sectors and processes mainly acting against sustained use are materialism, industrialization, capitalism, trends towards globalization; lack of effective state and political leadership, ineffective governance.

Equivocal sectors and processes—short-term policies, exacerbated by short-term time horizons of political parties and others, aimed at ameliorating the degradation of the environment at world, regional, national and local levels

Sectors promoting sustained use are effective state and political leadership and governance, supporters (governments, non-government organizations and individuals) of international pressure groups—Greenpeace, World Wildlife Fund (WWF), Soil Association, Free Trade, Forest Stewardship Council, Rainforest Alliance, Stakeholder Forum, charities, missions etc. Supporters of national and local pressure groups are supporters (subscribers) of educational and information journals—*National Geographical Magazine* etc.

Dryzek (2005) analysed eight discourses or views on environmental concerns that have been adapted in the box below:

Speth (2004) in his *Red Sky at Morning* suggests that past action has been by *The Incremental Approach* to solving the problems of the environment. This was, perhaps, the earliest attempt to find a solution and is typified by 'Green Peace' activities taking direct action to prevent pollution at sea through the sinking of disused oil rigs, or the dumping of toxic waste. Such activities are extremely useful and beneficial in highlighting aspects of problems, but they tend to concentrate on those with an element of the spectacular rather than those that are most damaging. The

Box 1

Eight Discourses on Environmental Concerns

- Survivalism—based on the concept of limits to growth set by finite natural resources; if growth is not limited then the ecosystem will collapse.
- The role of technology—the Promethean response—just as Prometheus stole fire from the Gods and increased the power of humans, so will advances in technology allow the expansion of resources to meet human growth and needs.
- Solutions through technocracy—administrative rationalism—leave it to the experts to impose solutions.
- Solutions through democracy—democratic pragmatism; the evolution of policy as a result of diverse views among equals.
- Solutions through the market—economic rationalism; assuming free competition and the fair pricing of all goods and services.
- Environmentally benign growth—sustainable development; assumes economic growth, environmental protection and sustainability are compatible.
- Ecological modernization of society and industry; governments, businesses, scientists, environmentalists all cooperate in the reconstruction of the capitalist political economy along environmentally sustainable lines.
- Green radicalism: mankind subject to nature; idealistic; mankind and its institutions must change to save the planet.

After Dryzek (2005).

approach is one that allows local situations to be the centre of interest and encourages local support.

The *Incremental Approach* or Domestic Agenda to tackle the problems of the environment of the 1970s is contrasted with a Global Agenda of 1980s. The former is characterised as a response to an acute problem, affecting us here and now, currently visible and scientifically understood. Whereas problems of a global nature tend to be chronic, affecting other people in faraway lands, in the future, remote and difficult to perceive, complex and difficult to understand. The pollution with CFC gases and the development of the hole in the ozone layer typifies a domestic problem apparently successfully dealt with through the Montreal agreement, whereas failure of both the Rio and Kyoto agreements typifies the more complex problems of a global nature.

The Incremental Approach to a domestic, highly focussed or local problem is a powerful tool to engender action but one that, to my mind, encourages fragmentation and obscures the interactions between the elements that are of much greater importance than the sum of the parts. Speth refers to the whole of such initiatives as the *Jazz scenario*—unscripted, voluntary, uncoordinated, improvised initiatives. Currently, in the UK, great emphasis is being directed by the government on reducing and recycling waste materials—both from homes and industry. Unfortunately, the motivation for this comes from comparison between countries within the European Economic Community— rather than from a holistic study of consumption, pollution and waste by community member states. Many will argue that the whole is too complex for mass understanding and that, by focussing on a section, one illuminates, at least in part, the whole. Such a focus has been widely adopted as being the only practicable approach to a solution, and one often adopted by governments with a short time horizon (up to the next election). This scenario contrasts with that termed *Geopolity*—in which people turn to international bodies and conventions to solve the problems of the environmental and society.

What is the part to be played by the demand for the sustained use of natural resources, the protection of the environment and the conservation of the ecology and biodiversity of the planet by a relatively uncaring and passive society? Why is society so unconcerned? What is the rate of change in outlook and effective action relative to the scale of the problem? Can society reverse the trend of unsustainable use of naturally occurring resources, reduction of degradation of the land and water resources, reduction of pollution and reversal of global warming? How much is due to ignorance, inertia, selfishness? Speth argues powerfully that all is not lost.

If mankind and the world as we know it are to co-exist, we must believe that the demands and efforts of those concerned, will slow down the rate of degradation in the environment, but such a

Box 2

- "Only the rich can afford to economise",
- "A hungry person sees an animal as a potential meal—not as something to be conserved",
- "Persons facing starvation cannot look beyond tomorrow",
- Only the rich can afford altruism and long-term viewpoints.

belief will not, by itself, reverse the current trends in population growth, materialism and bad governance that are the root causes of the degradation, or improve equity in the distribution of the costs and benefits of development. Unsustainable use of resources is often a reflection of the struggle to survive, poverty, ignorance and short-term views. This, however, is only a small part of the story as the industrial societies of the world are responsible for a major part of the pollution through the use of fossil fuels for industry and transport.

The contrast in rate of growth of the population and the use per head of energy and pollution per head between the subsistence and developed economies will have an increasing impact in the near future as subsistence economies in countries with large populations—such as Brazil, India and China—become part of the global market. The Earth Summit 2002 made little effective, practical progress towards solving the problem of how to prevent the situation worsening, as developing countries strive to match developments elsewhere. It may have raised awareness in some corners with its focus on programmes, indicators and monitoring, but there were few interconnections between the separate deliberations and resolutions on population, poverty, health, trade, water, energy, pollution, industrialization, globalization, urbanization etc. No nation or government seems able or willing to limit their aspirations and retreat from the spiral of materialism, industrialization, globalization, urbanization and associated trends that contribute to the degradation of the planet's environment. The Report of the Commission for Africa[11] has two pages out of over 400 allocated to considerations of sustainability; its recommendations are bland.

Currently, in Europe, much attention is being paid by governments to stabilising emissions from the use of fossil fuels, trading carbon emissions and recycling as a means of reducing demand on resources. There are some moves towards promoting local food markets rather than centralizing production, and fostering fair trade to spread the benefits with greater equity. However, these are barely more than palliative measures and they will do little to reduce the severity of the impact of mankind on the environment. Unless we find means of dispensing with the use of fossil fuels, then the damage from industrialization, transport and materialism will continue, and be exacerbated as the increasing aspirations and demands from the population of the currently less developed countries are met.

Will Man's characters of inventiveness, violence, elitism mimicry and selfishness turn out to be fatal faults, and those of intelligence, understanding, cooperation, altruism etc., be insufficiently developed and practised to ameliorate the costs of development? The impact of altruistic actions is demonstrated whenever there is a disaster such as the tsunami that devastated Sri Lanka, earthquakes, the floods of Mumbai and New Orleans etc.,—but it is less in evidence in the course of the day-to-day activities of mankind.

Speth outlines eight areas where transitions from the present to a sustainable situation are required to save the planet:

Box 3

- A stable or diminishing population
- Freedom from poverty
- Environmentally benign technologies
- Sustainable consumption
- Environmentally honest prices
- Developing good governance
- Development of knowledge and learning about the world system
- Changing culture and consciousness

After Speth (2004).

It is true that the rates of increase in some sectors of the global population are decreasing and will change with improved health, social security etc., but as fast as this happens, problems derived from increased longevity arise. Similarly, as poverty is relieved the aspirations of citizens increase and the gap between the extremes of society does not narrow.

Environmentally benign technologies are being developed and there is little doubt that society has learned some lessons from the past, nevertheless the short-term rewards from cutting corners remain attractive and unless the human race does change its culture, we are afraid that disasters will occur. Sustainable consumption has strong links with stabilising the global population and seems a goal that will be difficult to reach. There is plenty of evidence that people will change their consumption patterns to meet environmental standards—but only so long as their own convenience is not impaired. People will change to eco-friendly fuels and engines but will not give up their cars.

Speth's call for environmentally honest prices arouses my idealistic instincts and enthusiasms, but the problems involved centre on how to set these prices and how to ensure they are paid. What is the value of sustainable water supplies or an endangered species, and who pays the price? Good governance is essential to ensure that those who incur the costs are given compensation, and those who benefit pay the costs. It is of little benefit to those dependent on the forest and the forest dwellers for altruistic individuals supporting the conservation of the tropical rain forest and endeavouring to prevent unsustainable logging by renting the land, if the benefits are sequestered by elites. The problems of governance go far beyond the bounds of the forest, reach across continents, involve weapons dealing and other aspects of international trade and involve the very essence of nationalism, beliefs and cultures. Hence the importance of the last two transitions—stressing the need to learn and apply knowledge about the world system wisely and changing cultures and becoming conscious of global realities. Are these changes a real possibility or only a counsel of perfection?

WHAT ARE THE ROLES OF RESEARCH AND EDUCATION?

There is information available and more people are aware of many of the problems associated with unsustainable practices and use of resources. Formal education from primary to higher levels, as well as informal education through television, the worldwide web etc., now have a strong content of conservation, sustainability and the dangers from global warming. Nevertheless, very many state authorities, local authorities and individuals maintain an equivocal position—advocating sustainability but practising unsustainable levels of consumption. Some, especially those at the subsistence level, have no alternative, but others award material comforts and current consumption a higher priority than sustainable consumption and development.

If the global society is incapable of looking ahead and unable to curb its appetite for consumption, then one alternative is to 'hope', or 'have faith' that technology will unlock as yet unused resources and deal effectively with pollution. This standpoint has to be matched with and supported by effective research—into 'clean' energy, reduction in the production of pollutants and their more effective removal.

Such research will need the support of effective programmes of education, population control etc., to curb demand and, possibly, to reverse the trends in globalization.

CONCLUSIONS

There has been a radical change in outlook during last five decades. Unfortunately, this has been accompanied by acceleration in the rate of pollution and unsustainable development. Now the problems are well known and widely discussed in most countries—even if the causes are still debated and the solutions remain obscure and uncertain. Will people heed the lessons of the past and the costs of ignoring the obvious?

There are numerous signs that mankind is changing, but also examples that suggest that 'the devil is still active'. How long have we got to introduce effective changes in our behaviour? There is no doubt about the change in the level of appreciation of the dangers in the current situation—global warming is constantly in the news, environmental impact assessments (for what they are worth) are standard in many countries, poverty and disease are common subjects of discussion etc. Of Speth's eight transitions I see little cause for optimism in five-control of population, government-though there does seem hope in the introduction of benign technologies, developing knowledge and learning about the world system and in changing peoples' culture and consciousness.

If the levels of pollution can be reversed, then the flexibility of the global system may well cope, and the originality and technical ability of mankind may meet the challenge with success. Whether there will be sufficient time to correct the imbalance introduced into the system is in doubt. Optimists will believe that all will be well; others will be less certain.

NOTES

1. The main source of information was the Population Reference Bureau's 2005 World Population Data Sheet. www.prb.org
2. World energy projections, resources, consumption and CO_2 emissions www. combusem.com and Energy Information Administration, www.eia.doe.gov
3. National Climatic Data Center http://www.ncdc.gov/oa/climate/severeweather/extremes html and Easterling, D.R., J.J. Evans, P. Ya Groisman, T.R. Karl, K.E. Kunkel and P. Ambenio (2000). *Bull Am. Meteorological Soc.* 81, 417–425.
4. Ditto.
5. The Population Institute www.populationinstitute.org
6. World Water Councils. www.worldwatercouncil.org
7. www.combusem.com
8. During a meeting in Amman in 2003, the Club of Rome announced the foundation of the Trans Mediterranean Renewable Energy Cooperation-TREC. TREC has been exploiting the tremendous solar energy potentials available in the MENA region. The main aim of this project is to develop a master plan for the rapid and large scale introduction of solar energy as a clean energy source with permanently sinking coasts, unlimited and inexhaustible reserves and significant benefits for the MENA countries.
9. World Energy Outlook 2004.

10. The Population Institute www.populationinstitute.org
11. "Our Common Interest". Report of the Commission for Africa (2005).

REFERENCES

Angelsen, Arild and Sven Wunder (2006) "Poverty and Inequality: Economic Growth is Better than its Reputation". In Banik, Dan (ed.), *Poverty, Politics and Development: Interdisciplinary Perspectives,* Fagbokforlaget, Bergen.

D. Orrell Gaiia. *Theory: Science of the Living Earth.*

Dawkins, Richard (1979) *The Selfish Gene,* Oxford University Press.

Dryzek, John F. (2005) *The Politics Of The Earth* (2nd ed.) Oxford University Press.

Fay, C. and G. Michon (2005) "Redressing Forestry Hegemony Hen a Forestry Regulatory Framework is Best Replaced by an Agrarian One", *Forests, Trees and Livelihoods*, 15 (2), 193–210.

Lovelock, J.E. (1979) *Gaia: A New Look at Life on Earth,* Oxford University Press.

Maslow, Abrahan H. (1943) "A Theory of Human Motivation", *Psychological Review* 50 (4), 370–96.

Speth, J.G. (2004) *Red Sky at Morning—America and the Crisis of the Global Environments—A Citizen's Agenda for Action*, Yale University Press, New Haven and London.

Westoby J. (1962) "The Role of Forest Industries in the Attack on Economic Underdevelopment" in *The State of Food and Agriculture* FAO, Rome.

Wiley, L.A (2004) "Can We Own the Forest? Looking at the Changing Tenure Environment for Community Forestry in Africa", *Forests, Trees and Livelihoods*, 14 (2, 3 and 4), 217–228.

2

Limiting Resources and Consumptive Behaviour

From Social Dilemmas to Ethical Self

PARUL RISHI

Humanity must live within the capacity of the earth. There is no other rational option in the longer run.

IUCN/UNEP/WWF, 1991

Limiting resources and increasing consumerism on the part of people are the fundamental issues threatening the globe. Increasing population growth, land erosion, declining food harvests, exhaustion of mineral deposits, shrinking petroleum reserves, air pollution, failing water supplies, dangerous water quality, acid rain, solid waste disposal, litter control, toxic wastes, and nuclear wastes are a few such examples. "In historical terms, the casual connections between economic growth, increases in personal consumption and increasing environmental damage have remained steady in industrial economies for the last two centuries ... . What awaits on the horizon are limits and constraints never before encountered, from a dizzying height of economic prosperity never before attained" (Goldblatt, 1996).

Over the last two hundred years, the combination of capitalism and industrialization has resulted in unprecedented pressure on the physical, chemical and biological systems that support life on the earth. Dramatic changes are occurring in numbers and geographical systems that support life on our planet. Dramatic changes are occurring in the ranges of animal and plant species; the climate system is becoming more unstable; increasing numbers of toxins permeate air, soil and water systems. Scientific evidence suggests that environmental limits are being reached—and breached—with uncertain outcomes for the future well-being of people and nature (Burgess et al., 2003).

Inhumanity Towards Environment

However, a bright ray of light is that after decades of public debate and various summits, the environmental issues are now familiar to all of us. We are quite aware that on the one hand, some vital natural resources are being depleted quickly, while on the other, there is an increase of undesirable by-products from our use of the very same natural resources (automobile fumes and litter). Various earth summits have remained dominated by debate about how best to manage the accelerating impacts of human activities in destabilizing and degrading environmental systems. The pressures, at a global scale, have increased dramatically in the second half of the twentieth century: Between 1950 and 1990, the world's human population has more than doubled (from 2.6 billion to 5.3 billion).... Water use nearly tripled, fish consumption grew 4.4 times, and energy use quintupled.... the use of chemical fertilizers increased roughly tenfold; world production of organic chemicals, major sources of air and water pollution, rose twentyfold. On an average, resource use per person has nearly tripled between 1950 and 1990. This growth has resulted in roughly a sixfold increase in human impact on the global environment during the four decades (Corson, 1994).

In response to the growing scientific evidence of human impacts on environmental systems, most social scientific research and policy development has focused on production issues such as the role of new technologies in widening the range and usefulness of material resources; the significance of capital formation and structural changes in economies in stimulating demand; and the wealthy countries of the north were identified as bearing most responsibility for the depletion and degradation of environmental resources.

The majority of productive resources have come from the poorer countries of the south, whilst their consumption is concentrated in the rich industrialized nations of the north. It is also clear that the huge investment in the science of global environmental change has deflected political (and scientific) attention away from local environmental problems, such as soil erosion/loss of fertility/destruction of habitat, which lead to further impoverishment of the poor people in many parts of the world (Agnew and Warren, 1996). The social practices and environmental consequences of contemporary consumerism were placed on the policy agenda following the Brundtland Report. Accordingly, the 1992 Earth Summit was called for "national policies and strategies to encourage changes in consumption patterns" (UNCED 1993: 64).

Questions of whether and how to ameliorate the impact of human activities on nature have surfaced periodically over the last 150 years (Thomas, 1955; Worster, 1994). Current debate was stimulated by the first UN Environment Conference is Stockholm, in 1972. The context was one in which the depletion of natural resources seemed to be setting limits to the continued growth of industrial economies (Meadows et al., 1972). In the 20 years, between Stockholm and Rio in 1992, debate had shifted to focus on the inability of environmental systems to cope with the present (and predicted) levels of economic activity. The 1992 Earth Summit addressed global human impacts on climate and habitats, and how to achieve more equitable but sustainable growth within bio-physical limits. Non-governmental organizations (NGOs) began to work in partnership with national governments in the north to promote the concept of 'environmentally-friendly lifestyles': first, to address production issues in poorer countries through the promotion of fairly traded goods, and later to reduce the impacts of consumption practices on environmental systems.

As environmental settings and natural resources are prone to sustainable/unsustainable consumerist behaviour, it is the social psyche and behaviour which needs to be addressed to bring the society out of the trap of common dilemmas and take it towards development of the ethical self for global environmental conservation. This chapter considers the psychological processes involved in managing limited environmental resources and places emphasis on those social and behavioural issues that have evolved out of management of natural environments and are linked to irresponsible consumptive behaviour.

Consumptive Behaviour

Consumptive practice is fundamentally political. Large organizations, public authorities, the military and commercial companies, are largely responsible for the bulk of human environment transactions which lead to damage to bio-physical systems, as Stern et al. (1997) acknowledge. But the focus of attention in terms of achieving changing consumption practices has remained firmly fixed on individual consumers and their domestic consumption. Accepting responsibility for one's actions puts pressure on individuals to make changes in their everyday lives and spaces, and to endorse political campaigns for increased regulations, standards and controls.

As members of a society, we extract, refine, use, and dispose of many natural resources. As individuals, our role in this process is primarily as users and disposers of processed natural resources. Few of us would seriously advocate an absolute halt to this process, except perhaps in the case of such nearly depleted resources as endangered species. This implies that we will continue to use and dispose of or recycle most resources. The term management here refers to the rate and quality of the individual's role in the cycle. The rate of this process is often quantifiable, the quality of the process is often debatable. For example, if a resident of a drought-threatened town decides to water a newly seeded lawn, it would not be difficult, using a water metre, to identify the quantity of the resource used. But whether the resident is correct to use the water is not easily decided (Gifford, 2002).

Much of the environmental management is done by governments, corporations, and other organizations. Such management is performed at a macro or societal level. However, resource management behaviour of individuals and small groups at the micro-level is equally important. As individuals, we are faced with our own resource management problems. In varying degrees, we monitor our personal use of resources, observe the effects that our usage has on the environment, and are aware of the usage patterns of other individuals. Our resource management decisions are based on all three of these considerations, and more. The crucial aspect of micro-level resource management is that it sums up to the macro level in mysterious, irrational, yet all important ways. Therefore, the interplay between an individual's resource management and what the individual perceives others' resource management to be, is an issue of discussion.

Research by psychologists, economists and social scientists focuses on the individual, with a particular emphasis on the importance of cognitive processes in consumerist choices and other environmentally-relevant behaviours. The range of new research that is drawing upon sociology, cultural geography, anthropology, and science and technology studies to understand consumption discourses and practices. An emerging research agenda is identified, based around the ecological-socio-technological networks that bind producers and consumers, plants and animals, natural and physical systems into relationships across space and through time.

MANAGING THE COMMONS

Managing the commons is a Herculean task in today's world when everyone is running after a consumerist behaviour. The basic assumption behind establishing the commons was the supply of resource should be able to meet the demands of the community in a sustainable way. Unfortunately, the earlier economists like Adam Smith were not able to see and propagate the limits of the resources, and people were free to exploit the resources to the extent possible. According to them exploitation of resource, its processes and usage ultimately leads to development of society.

An early nineteenth century economist, William Lloyd, was one of the first to see a problem with Smith's logic. The essential difference between the ideas of Smith and Lloyd was the recognition that many resources are limited. According to the biologist Hardin, in a desirable but limited commons, individuals act in self-interest. Hardin asserts that a process called the tragedy of the commons begins. "Each man is locked into a system that compels him to increase his herd without limit—in a world that is limited. Ruin is the destination toward which all men rush, each pursuing his own best interest ..." (Hardin, 1968). The similar fundamental ideas apply to water shortage, food supplies pollution, over population, energy and many other resources.

Thus a commons is any desirable resource held jointly by a group of individuals, from an office lottery pool to the oceans. Some of these resources renew themselves relatively quickly (grass for grazing or river water for electric power), others not so quickly (tree used for lumber) and some very slowly or not at all (oil, endangered species). Hardin and like-minded thinkers believe that it is difficult or impossible for us to reverse our tragic path toward ruin. The only thing that can save us, according to Hardin, is a "fundamental extension of morality, which means, mutual coercion, mutually agreed-upon." That is, society must agree, through laws and regulations, to severely limit individual freedom to exploit the commons. Hardin does not seem optimistic that society can accomplish this limitation because the individual's freedom to exploit resources is so deeply valued by many members of society.

SELF VS PUBLIC INTEREST

Common dilemma is a difficult choice whether to act in self-interest or in the public interest. The following lines indicate how people themselves decreate the nature by acting in the self-interest

In the manner of Shakespeare's Hamlet we may state "to go ahead or to cooperate, that is the question". When commons are in abundance, people have the choice to cooperate for public interest and also to exploit for self-interest, but when the situation is reverse, it is a loose-loose situation. Still, there are individual differences in self vs public interest choices. Some people will go out of the way to conserve the resources and managing the limited resources in a responsible manner. While, the others may simply remain unconcerned about what is happening to the environment.

Self-interest behaviours, like easy disposal of garbage, continuously switching on electricity, filling the pools in spite of water shortage, may appear to be quite rewarding and convenient in the short-term, and public serving acts, especially with the limited resources appear to be expensive but such self-serving behaviours put a great load on environment and the commons and environmentally responsible behaviour demands, on the part of the people, to behave in a public interest way.

Why People Behave in Self-interest?

The major reason is the limited resources. Everyone thinks that if he is not exploiting the resources, the others may exploit it and ultimately, the resource will be consumed with no benefit to them. So, why not they also exploit to benefit themselves at least proportionately to what others are getting. For example, there is a common woodland near few villages. Every village knows that if they will exploit it unsustainably by grazing their animals or illicit felling, the woodland will be lost because of their consumerist behaviour, but because they fear, that the fate of woodland is going to be same, if other villages exploit it, and they will remain not benefited by it, so let everyone consume it. This tendency is known as 'defection'. People who defect, are cognitively aware about the limited resource and effects of their irresponsible consumerist behaviour, but cannot actively motivate others to behave in a public interest way and hence, find it easy to be the part of them in consuming the resource.

Psychologists view this problem of resource management from two different perspectives. A particular group may be involved through their interest in choice, decision making and rationality, while the other group may become interested due to

Box 1

Two persons are involved in illicit felling in the forest. They are apprehended by the village head who knows that they are involved in felling but do not have sufficient evidence to convict them. The local court puts them in separate prison rooms and orders to make an identical proposition to each. If either prisoner confesses, the other will be sent to jail for 4 years. If both confess, they both will go to jail for 2 years. If neither confesses, they both will go to jail for 6 months on some minor charge. It was informed to both the prisoners that the similar proposition has been made to the other one too and they were left in separate rooms to think.

direct environmental concern. It can be related to the famous game of Prisoner's Dilemma which goes as in Box 2a.

Gifford (2002) indicates the similarity between prisoner's dilemma and commons' dilemma. At one side, there are two participants in prison and the issue is interpersonal trust. However, on the other side, there are many participants in resource use dilemmas and the issue is shrinking resources. Decision making for free resource use will be appropriate only when similar decision is not being taken by any other individual, which is next to impossible. Dawes (1973) used the term common dilemma to refer to the overuse of natural resources as a result of conflict between individual and group interest.

Social trap (Platt, 1973) is explained as a situation in which people succumb to the immediate advantage without taking into consideration the built-in and gradual cost which eventually becomes large and people are in the temporal trap. Dawes (1980) suggested the term 'social dilemma' to encompass common dilemma, social trap and even Prisoner's dilemma.

Cooperation vs Defection

Consumptive behaviour depends on the three major variables—the resource itself, the people who consume and the rules of the game. In regard to resource, as the value of resource increases, the rate of cooperation decreases (Kelley et al., 1965). Besides, if the resource is almost gone, or degraded, the cooperation is more than if the resource appears to be plentiful (Rubinstein et al., 1975). Regarding people or participants, cooperation declines with increase in the number of people/groups who are dependent on the resource. Accepting responsibility for one's actions puts pressure on individuals to make changes in their everyday lives (McCormick, 1995). All consumptive practices do not damage the environment. Only in appropriate consumption, which may have negative consequences for consumers as well as the resource, is to be identified and checked.

SOCIAL DILEMMAS: THEORETICAL PERSPECTIVES

Social dilemmas can be explained under five theoretical perspectives as per the interpretations of Julian Edney (1981) and Dawes (1980).

Tragic Choice

The first one is tragic choice theory (TC). We all would prefer a world without social dilemmas where all desirable resources are accessible in plenty; to each one of us. However, the fact is that in spite of the availability of resources to the same extent, the distribution is not equal. Calabresi and Bobbitt (1978) have labelled it as TC theory. The postulate that inequality in resource allocation results in scarcity and suffering which is tragic choice and difficult to change.

The theory holds that equal allocation is possible only in case of more powerful central authority than most of us would tolerate. The theory further indicates that conscious decisions are taken to favour the cause of a selected few as against the common public. For example, producing electricity or good or other resources less than its maximum capacity to create scarcity and stabilize prices. Besides, insufficient distribution is another example of the society not to value every person equally. So, unthoughtful consumerist behaviour, as per TC, is not always the result of limiting resources. It is primarily the result of self-interested choices of a few as justification for getting ahead of others by over using the commons. A more honest and empathetic attitude on the part of the selected few people or countries is desirable to realise the relationship between their consumerist behaviour and the suffering of the rest of the people or countries.

Game Theory

Game theorists try to make sense of the choices we make by reducing that choice to the simplest one possible. If a simulation is properly constructed, the experimenter can see very clearly what an individual means by each choice. Most simulations assume that everyone starts out equally and that the rational choice is to defect, hoping most others cooperate. Of course, this 'power-tripping' is dangerous, it leads to disaster when others try the same strategy. Research indicates that defectors are also unpopular, although, in everyday life, defecting strategies are frequently defended as being realistic or even as beneficial to society. Cooperation, on the other hand, is more often socially acceptable, but it is also dangerous. Others may take advantage of you.

The essence of the social dilemma from the game-theory point of view is trust. With trust, participants maximize their joint outcomes. Without trust, defection makes the most sense, yet it will almost certainly lead to the worst outcome for all when repeated choices are made.

Social Trap Theory

It is based on John Platt's framework of reinforcement. Many of us want to reward ourselves and if the reinforcement structure of social dilemma is examined to rearrange the timing and value of reinforcement, the negative outcomes can be reversed. The primary reason for mismanagement of commons, as per social trap theory, is that reinforcement for public interest behaviours are not only smaller than self-interest behaviours, but are not contingent on the positive behaviour as such because of their occurrence long after public spirited act.

For example, the village person who conserve the forest or the common man who always tries to conserve the oil by using public transport does not have any recognition for his/her public spirited act, whether timely or even delayed. Similarly, the person littering of public places is not immediately and firmly informed of his/her anti-social act and advised for its rectification. So, the faulty positive/negative reinforcement framework associated with public interest behaviour causes people to be in the social trap for self-interest behaviour promoting further irrational consumption of resources or environmental undesirable behaviour.

Equity Theory

Equity theory focuses on cognitive processes involved in sharing resources. The theory indicates that equity is the ratio of net rewards (outputs) to inputs (effort, skills, education etc.). For example, doctors earn handsome income but as the general public perceives their profession as life-saving, besides knowing the amount of money and hard work put in by them to reach that level, their unequally higher share of income in the society is acceptable. Equity theory claims that people are willing to tolerate resource inequalities as long as those who have additional resources have worked harder to achieve the extra share. So what is desirable is the equality of input-output ratio.

The theory further asserts that we have the ability to distort outcomes so that they seem equitable to us. This theory does not take into account the limiting resources and over consumerist behaviour of too many individuals simultaneously targeting the resource exploitation. It may be rather frightening if people compete to defect for the larger share of resources. However, for promoting pro-environmental behaviour, equity theory can be used to modify the cognitions of communities as a whole with the proposition that if one group conserves resources (input through positive attitude and effort) the output will be sustained a availability of resource and felicitation by government.

However, the other group not conserving the resources will be refrained from these positive outcomes. On the other hand, if outcomes are equal for both the groups conserving or exploiting the resources or the exploiting group is getting more positive outcomes, it will lead to inequity which may even cause destruction of resource by both the groups.

Limiting Processing Theory

Gifford (2002) states that much of the socially/environmentally undesirable behaviour is the outcome of non-rationality on the part of users. This non-rationality is not always an active choice of the people, but because of limited processing, the defection occurs. This defection is not perceived as defection by the people who are doing it owing to unawareness of its possible outcomes on commons.

For example, a poor villager in a developing country is going to forest for collecting fuelwood which he uses for cooking. She/he may not be aware of the cumulative effect of this behaviour on common resources in the years to come due to limited processing. It is again a social dilemma whether the fulfill the immediate basic needs or to think for future availability basic needs or to think for future availability of resource without the availability of immediate alternative for cooking. Some of these dilemmas are simple to handle through education while others may be true dilemmas with no immediate solutions.

So people act selfishly not because they are evil, but because the dangers of defection do not occur to them in a short-term. The everyday examples of such dilemmas may be pollution, energy conservation, recycling, pesticide use etc., which are more described as a behavioural problem than technical/environmental issues. Increased awareness of consequences of overusing resources of limited supply may be one answer to them.

COGNITIVE AND MOTIVATIONAL PERSPECTIVES

A variety of social scientific disciplines has focused on the individual consumer and the cognitive and motivational bases of their behaviour as it relates to environmental concerns (Neuman, 1986; Oskamp et al., 1991; de Young, 1993). To what extent are individuals changing aspects of their behaviour for the sake of the environment?' Perhaps by buying organic products or refusing extra packaging; modifying household regimes by turning down the central heating or only using the washing machines with a full load; choosing not to use the car to travel to work.

Cognitive perspectives construe the problem as one of individual (and social) learning—that is how, over time, understanding of the necessity for behavioural change is achieved, and then, critically, implemented. Developmental theories underpin micro studies of individuals' environmental attitudes and behaviour, as Finger's (1994) concise summary of the 'dominant' environmental attitude and behaviour framework shows schemata through which complex information is assimilated into workable cognitive frameworks; social psychology is able to explain how individuals' value systems reflect bundles of more or less consistent sets of attitudes and beliefs, including important social norms; while environmental educationalists engage in pedagogic practices to ensure that children acquire the information and knowledge necessary to engender environmental concern and commitment.

It assumes that as the individual acquires new information and knowledge about the impacts of human activity on bio-physical systems, whether through formal education at school, targeted information from government departments or information channels such as the mass media, environmental awareness will rise, and existing environmental values and attitudes will change. In this 'virtuous circle', more information will enhance the individual's willingness to reduce the environmental impacts of their activities through behavioural changes.

A substantial body of data exists which charts rise and fall in public awareness of environmental problems, changes in environmental attitudes and reported behavioural changes (Dunlap and Scarce, 1991; Ungar 1994; Worster, 1993; Dunlap, 1998). A key finding from studies in many different parts of the industrialized world is paradoxical. The remarkably rapid increase in public awareness of environmental issues and embracing of pro-environmental attitudes is coupled with virtually no substantive changes in behaviour at all. The failure to demonstrate any convincing correlation between environmental awareness, attitudes and behaviour, to say the least, is unfortunate. For policy, so far has been relying through reliance on cognitive models of behaviour.

The strategy for engendering more sustainable lifestyles is seen as a matter of providing further and better information, with the expectation that this will raise environmental awareness and concern; lead to more informed consumer choices; and change purchase decisions and/or routine, everyday behaviour. Pro-environmental behaviour is thus constructed as the outcome of a linear and ultimately rational process.

Although there is evidence to suggest that fear and anxiety are powerful motivators in sensitizing individuals to the risks associated with environmental changes, recent empirical studies show that such communications are failing to persuade individuals of the necessity of changing their behaviour (Harrison et al., 1996; Myers and Macnaghten, 1998; Blake, 1999). New research is focusing on specific consumption practices, on the assumption that different domains of environmentally significant consumption may well engender different attitudinal and behavioural responses. One recent example is a major study by two economic psychologists (Brandon and Lewis, 1999) who conducted a field study with 120 household in Bath, UK.

The evidence suggested that "financial considerations are of equal or even greater importance" in decisions to reduce energy consumption than environmental attitudes (Brandon and Lewis, 1999: 83). Overall, the results indicated that the feedback was significant in reducing energy consumption over the period.

CONCLUSION: TOWARDS THE ETHICAL SELF

The above discussion of limiting resources and consumerist behaviour causing social dilemmas takes us to the point where we have to rethink about in which direction we are moving and heading towards what? It is all a matter of choices. Whether we want to move in the direction where materialism taken over conservation of natural resources? Or we want to behave in a pro-environmental way. Conserving the limiting resources and modifying our consumerist behaviour is a way that suits to the societal requirements.

It is true that self-interest choice provides immediate reward, but we reflect on different 'self' at different points of time, like real self (deal self, rational self, material self) and above all the 'ethical self'. Today's scenario of limiting resources on the part of globe and unmindful consumerist behaviour, on the part of the people makes it desirable for each and every individual to guide his/her behaviour by 'Ethical Self' which is your conscience, always providing you advice based on morality and justice for the society and above all the planet earth. Only then, we can foresee the sustainable futures.

REFERENCES

Agnew, C. and Warren, A. (1996) "A Framework for Tackling Drought and Land Degradation", *Journal of Arid Environments*, 33, 309–20.

Blake, J. (1999), "Overcoming the 'Value-Action' Gap in Environmental Policy: Tensions Between National Policy and Local Experience", *Local Environment*, 4, 257–78.

Brandon, G. and A. Lewis (1999), "Reducing Household Energy Consumption: A Qualitative and Quantitative Field Study", *Journal of Environmental Psychology,* 19, 75–85.

Burgess, J., Bedford T., Hobson K., Hobson, Davies Gail and Harrison C. (2003), "(Un) Sustainable Consumption". In Frans Burkhout et al. (ed.), *Negotiating Environmental Change,* EE, Great Britain.

Calabresi, G. and Bobbit, P. (1978) *Tragic Choices, The Conflicts Society Confronts in the Allocation of Tragically Scarce Resources,* WW Norton, New York/London.

Corson, W.H. (1994) "Changing Course: An Outline of Strategies for a Sustainable Future, Futures", *The Journal of Forecasting and Planning,* 26 (2), 206–23.

Dawes, R.M. (1980) "Social Dilemmas". *Annual Review of Psychology,* 31, 169–193.

Dawes, R.M. (1973) "The Commons Dilemma Game: An N-person Mixed-motive Game with a Dominating Strategy for Defection", *Oregon Research Institute Research Bulletin,* 13, 1–12.

Dunlap, R.E. (1998), "Lay Perceptions of Global Risk: Public Views of Global Warming in Cross National Context", *International Sociology,* 13, 473–98.

Dunlap, R. and R. Scarce (1991), "The Polls-Poll Trends: Environmental Problems and Protection', *Public Opinion Quarterly,* 55, 651–72.

Edney, Julian J. (1981) "Paradoxes on the Commons: Scarcity and the Problem of Equality", *Journal of Community Psychology,* 9 (1), pp. 151–159.

Finger. M. (1994). "From Knowledge to Action? Exploring the Relationships Between Environmental Experiences, Learning and Behaviour", *Journal of Social Issues,* 50, 141–60.

Gifford, R. (2002) *Environmental Psychology: Principles and Practice,* Colville, WA, Optimal Books.

Goldblatt, D. (1996) *Social Theory and the Environment,* Cambridge, Polity Press.

Hardin, G. (1968) "The Tragedy of the Commons", *Science, 162,* 1243–1248.

Harrison, C.M., J. Burgess and P. Filius (1996), "Rationalising Environmental Responsibilities: A Comparison of Lay Publics in the UK and the Netherlands", *Global Environmental Change,* 6 (3), 215–34.

Kelley, H.H., Condry, J.C., Jr. Dahlke, A.E. and Hill, A.H. (1965). "Collective Behaviour in a Simulated Panic Situation". *Journal of Experimental Social Psychology,* 1, 20–54.

McCormick. J. (1995) *The Global Environmental Movement,* Chichester, Wiley (2nd ed.).

Meadows, D.H., D.L. Meadows, J. Randers and W. Behrens (1972) *Limits to Growth*, London, Earth Island Press.

Myers, G. and P. Macnaghten (1998), "Rhetorics of Environmental Sustainability: Commonplaces and Places", *Environment and Planning A*. 30, 333–53.

Neuman, K. (1986), "Personal Values and Commitment to Energy Conservation", *Environment and Behaviour,* 18, 53–74.

Oskamp, S., M. Harrington, T.C. Edwards, D.L. Sherwood, S.M. Okuda and D.C. Swanson (1991), "Factors Influencing Household Recycling Behaviour", *Environment and Behavior,* 23, 494–519.

Platt, J. (1973) "Social Traps". *American Psychologist,* 28, 641–651

Rubenstein, F.D. Watzke, G. Doktor, R.H. and Dana. (1975), "The Effect of Two Incentive Schemes Upon the Conservation of Shared Resource by Five Person Group". *Organizational Behaviour and Human Performance,* 13, 330–338.

Stern, Paul C., Dietz Thomas and Alcock, Robert (1997) "Consumption and Sustainable Development", *Science,* 276. (5319), pp. 1629–1632.

Thomas, Balogh (1955) *The Economics of Poverty,* London, Weidenfeld and Nicolson.

Ungar, S. (1994), "Apples and Oranges: Probing the Attitude-Behaviour Relationship for the Environment". *Canadian Review of Sociology and Anthropology,* 31, 288–304.

United Nations Conference on Environment and Development (UNCED) (1993) Agenda 21, Programme of Action for Sustainable Development, New York, UN Department of Public Information.

Worster, D. (1994) *Nature's Economy: A History of Ecological Ideas,* Cambridge, Cambridge University Press (2nd ed.).

3

Climate Change and Human Psyche

A Relook at Choice

PARUL RISHI

Humankind has not woven the web of life. We are but one thread within it. Whatever we do to the web, we do to ourselves. All things are bound together. All things connect.

Chief Seattle, 1855

Our environment is dynamic, constantly changing and evolving in response to stimuli. Yet in the 20th century, it became apparent that mankind was having an increasing effect on the planet's ecosystems and biogeochemical cycles, so much so that our activities are now causing environmental change which is overriding the natural dynamism of the earth. Yet despite the evidence of environmental problems such as biodiversity loss, land cover change observable from satellite images, records of climate change and many examples of pollution, we still pursue activities which perpetuate the problems. As the world's population increases, and the per capita consumption of natural resources increases, we will have an even greater effect on these environmental problems, exacerbating them further (Harris, 2004).

Human-environment interactions involve not only the question of resource use per person, but also our ability to understand the science of the environment, our ability to regulate our impact on the environment, our beliefs in the value of the environment, our attitudes to the future, particularly risk, and our ability to negotiate solutions at both the local and the global levels.

Climate change is one of the greatest environmental, social and economic threats facing the planet earth. The climate system is unequivocally warming, as is now evident from observations of increases in global average air and ocean temperatures, widespread melting of snow and ice, and rising global mean sea level. Changing climate is continuously warning the climatologists, scientists, economists, decision makers and many others since decades. A vast majority of credible climate scientists well over 95 per cent, opine that average temperature of the oceans, the land surface of planet and the lower atmosphere have been going up at an accelerating rate (Blakemore, 2005).

Climate change refers to a statistically significant variation in either the mean state of the climate or in its variability, persisting for an extended period (typically decades or longer). Climate change may be due to natural internal processes or external forcing, or to persistent anthropogenic changes in the composition of the atmosphere or in land use. According to IPCC (2007), potential causes of climate change are both natural and anthropogenic, including solar changes, volcanic eruptions, greenhouse gases (GHGs), land covers and aerosols.

Atmospheric concentrations of greenhouse gases have grown significantly. These gases tend to have a warming effect by absorbing infrared radiation from the earth's surface and then re-radiating it. Atmospheric concentrations of aerosols grow as well. Aerosols are microscopic airborne particles that, on balance, tend to have a cooling effect. Although locally the cooling effect due to aerosols may be large enough to offset the warming effect due to greenhouse gases, this does not hold globally. Taking into account the effects of both greenhouse gases and aerosols, human interference is projected to increase the mean temperature on earth. IPCC (2007) further states that anthropogenic emissions grew by 70 per cent between 1970 and 2004. As a result, every form of life on this planet is likely to be subjected to impact of human action through changes in the climate (Pachauri, 2004).

Only the earth has an atmosphere of the proper depth and chemical composition. About 30 per cent of incoming energy from the sun is reflected back into space while the rest reaches the earth, warming the air, oceans, and land, and maintaining an average surface temperature of about 15°C. The atmosphere carries out the critical function of maintaining life-sustaining conditions on the earth, in the following way: each day, energy from the sun (largely in the visible part of the spectrum, but also some in the ultraviolet and infrared portions) is absorbed by the land, seas, mountains etc. If all this energy were to be absorbed completely, the earth would gradually become hotter and hotter. But actually, the earth both absorbs and, simultaneously releases it in the form of infrared waves (which cannot be seen by our eyes but can be felt as heat, for example the heat that you can feel with your hands over a heated car engine). All this rising heat is not lost to space, but is partly absorbed by some gases present in very small (or trace) quantities in the atmosphere, called GHGs (greenhouse gases). The greenhouse gases (for example, carbon dioxide, methane, nitrous oxide, water vapour, ozone), re-emit some of this heat to the earth's surface. If they did not perform this useful function, most of the heat energy would escape, leaving the earth cold (about −18°C) and unfit to support life. An increase in the levels of GHGs could lead to greater warming, which, in turn, could have an impact on the world's climate, leading to the phenomenon known as climate change.

Since the industrial revolution, as the world became excessively dependant on fossil fuels for energy, emissions of the greenhouse gases (GHGs) have increased dramatically. The concentration of carbon dioxide in the atmosphere has gone up by 31 per cent since 1750. These gases trap heat, thus leading to 'global warming'. In the 20th century, the earth's temperature increased by about 0.6°C due to increasing concentration of GHGs in the atmosphere. It is expected to increase by another 1.4–5.8°C by the end of the 21st century. Rising temperatures will cause a series of reactions like rise in the sea level, causing inundation and increasing salinity in coastal areas, change in weather patterns that affect agricultural productivity, and floods creating favourable conditions for the spread of disease carrying vectors leading to epidemics.

THE ASIAN SCENARIO

Asia is the largest continent of the earth. The region faces formidable environmental and socio-economic challenges in its effort to protect valuable natural resources. Land and ecosystems are being degraded, threatening to undermine food security. In addition, water and air quality are deteriorating while continued increases in consumption and associated waste have contributed to the exponential growth in the region's existing environmental problems. Furthermore, the region is highly subject to natural hazards, such as the 2004 Indian Ocean tsunami, the 2005 Pakistan earthquake, and the 2006 landslides in the Philippines. There is evidence of prominent increases in the intensity and/or frequency of many extreme weather events such as heat waves, tropical cyclones, prolonged dry spells, intense rainfall, tornadoes, snow avalanches, thunderstorms, and severe dust storms in the region (Cruz et al., 2007). Impacts of such disasters range from hunger and susceptibility to disease, to loss of income and livelihoods, affecting human survival and well-being. For example the extreme weather events in China during 2006 included major storms and flooding in the east and south, as well as heat and drought in central, western and north-eastern regions, killing more that 2,700 people and causing US$ 20 billion in damages.

Many environmental and developmental problems in Asia will be exacerbated by climate change. Under climate change, predicted rainfall increases over most of Asia, particularly during the summer monsoon, could increase flood-prone areas in East Asia, South Asia and South-east Asia. In Central and South Asia, crop yields are predicted to fall by up to 30 per cent, creating a very high risk of hunger in several countries. Global warming is causing the glaciers melt in the Himalayas. In the shorter term, this means shortage of fresh water, increased risk of flooding, erosion, mudslides and GLOF in Nepal, Bangladesh, Pakistan, and north India during the wet season. In the longer term, global warming could lead to a rise in the snowline and disappearance of many glaciers causing serious impacts on the populations relying on the seven main rivers in Asia fed by melt water from the Himalayas. Throughout Asia one billion people could face water shortage leading to drought and land degradation by the 2050s (Christensen et al., 2007; Cruz et al., 2007).

In Asia, the principal impacts of climate change on health will be in the form of epidemics like malaria, dengue, and other vector-borne diseases (Martens et al., 1999). Projected sea level rise would flood the residence of millions of people living in the low lying areas of South, South-east and East Asia such as in Viet Nam, Bangladesh, India and China (Wassmann et al., 2004; Stern 2006; Cruz et al., 2007) and 30 per cent of coral reefs would be lost in the next 10 years (Cruz et al., 2007). The loss may be as high as 88 per cent (59% of global) in the next 30 years (Sheppard 2003; Wilkinson 2004). The Third Assessment Report (TAR) of IPCC rightly states that the impacts of climate change will fall disproportionately upon developing countries, thereby exacerbate inequities in health status and access to adequate food, clean water and other resources. The effects of such changes are already being felt. Some 262 million people have been adversely affected by climate disasters between 2000 and 2004 and 98 per cent of these are in developing countries (Yechuri, *Hindustan Times*, 24th January 2008).

WHY DO WE FAIL?

Why are such problems so difficult to resolve? The possible broad reasons are threefold. First reason is the complexity of environmental problems as we are dealing with multiple interrelated dynamic systems within which the feedback mechanisms occur (Harris, 2004). Secondly, the involvement of different stakeholders involved in the dynamics and solutions to such problems is also important. It is quite difficult to coordinate such stakeholders for the common interest. Thirdly, and most importantly, resolving global environmental issues will require changes in human consumption which requires changing the lifestyle through personal commitment, which not everyone is willing to do. Human-climate interactions involve not only the issues of managing one's own personal emissions but also the ability to understand the science of climate, human ability to regulate its impact on climate, one's beliefs in the impacts of changing climate, one's attitudes to the future risk and one's ability to negotiate solutions at both local and the global level. "The science of environment is socially and politically situated rather than unambiguous or reparable from the subjective location of human perception" (Stott and Sullivan, 2000).

IMPACTS OF CLIMATE CHANGE

The developing countries, on the other hand, have taken the road to growth and development very recently. In countries like India, emissions have started growing but their per capita emissions are still significantly lower than that of industrialized countries. The difference in emissions between industrialized and a developing country is even starker when per capita emissions are taken into account. For instance, the emission of one US citizen equaled that of 20 Indians (*HT*, 24th January 2008). Ironically, even though developing countries' contribution to global warming is minimal, they stand to face most of its adverse effects. They are twice as vulnerable to climate change as compared to industrialized countries. For instance, while agriculture in developing countries will be negatively affected, a few industrialized countries will actually benefit from increase in crop productivity as climate changes. Moreover, developing countries have limited resources to cope with and adapt to changes which global warming will bring in its wake. But the industrial world has failed in its responsibility to do something urgently to minimize the impacts of climate change. The United Nations Framework Convention on Climate Change (UNFCCC) agreed to in 1992 and the 1997 Kyoto Protocol, the two international agreements to deal with global warming, have not yet succeeded in reducing GHG emissions. The conference, hosted by the Government of Indonesia, took place at the Bali International Convention Centre and brought together more than 10,000 participants, including representatives of over 180 countries together with observers from inter-governmental and non-governmental organizations and the media. Parties have recognized the urgency of action on climate change and have now provided the political response to what scientists have been telling us is needed. The conference culminated in the adoption of the Bali roadmap, which charts the course for a new negotiating process to be concluded by 2009 that will ultimately lead to a post-2012 international agreement on climate change. Decisions on technology transfer and on reducing emissions from deforestation were taken to represent various tracks that are essential to achieving a secure climate future. A special adaptation fund has been created to support the vulnerable countries from the impact of climate change (UNFCCC, 2007). However, Sitaram Yechuri (2008) writes in an editorial in the *Hindustan Times* that in view of 20 tonnes of emissions per capita in US, reducing 80 per cent means three tonne per capita by 2050. Reduction of 20 per cent in India would mean 0.8 tonne per capita. Thus India will be bearing the burden of industrialized countries which is not justified. He further states that per capita emissions must be the basis for solutions which should be stressed upon in 2009 roadmap.

HUMAN FACE OF CLIMATE CHANGE

Humans have long held the capacity to alter the weathering systems advertantly or inadvertently. The balance of evidence suggests that there is a discernible human influence on the global climate. This conclusion was drawn by the Intergovernmental Panel on Climate Change (IPCC, 2007). The IPCC was established ten years ago to assess available scientific information on climate change, to estimate the environmental and socio-economic impacts of climate change, and to formulate response strategies.

According to the IPCC (2007) presentation in CoP–13 summit at Bali in December, 2007, the composition of the earth's atmosphere is continuously changing due to anthropogenic activities at a faster pace. The sheer number of people on the planet, their demand for (predominantly fossil-carbon) energy and the waste by-products of consumerist societies, have now, however, created the potential for humans to alter the global characteristics of the physical planetary system that delivers weather and climate to individual people and places. Whereas in the past, climates shaped and constrained societies, now global society itself can modify global and regional climate. The dynamics of human-climate interface has been described in Figure 3.1.

As it appears from the expression of reduction targets in CO_2 equivalents, one of the most important anthropogenic greenhouse gas is CO_2 (Fransen and Janssen, 1998; IPCC, 2007). Emissions of CO_2 are mainly due to the combustion of fossil fuels for the

Figure 3.1 Dynamics of Human Climate Interface and Adaptation

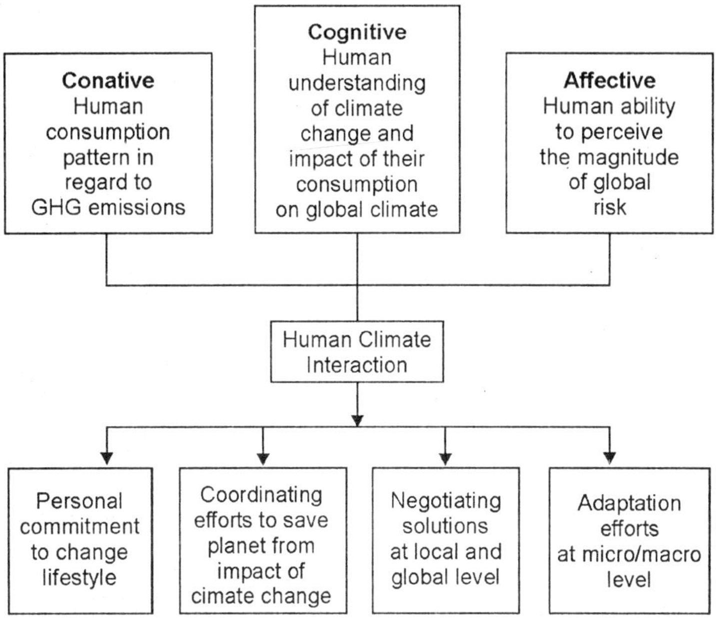

generation of energy. The global demand for fossil fuel has grown for almost two centuries and is expected to continue to grow at least through the first half of the next century. Unless measures are taken before long atmospheric concentration of CO_2 will also continue to grow. Significant reduction in CO_2 emissions can be achieved by replacing current technology with more energy efficient technology and by switching to low-carbon fossil fuels and non-fossil fuels. In the longer term renewable energy sources such as solar, wind, and biomass technologies could meet a considerable part of the world's energy demand. Extensive research is being undertaken to explore various said and unsaid dimensions of changing climate to have a more controlled approach over nature to sustain it for the future generations. However widespread motivational and awareness generating efforts are required in this regard.

NEED FOR BEHAVIOURAL SCIENTISTS

In spite of all such efforts, 'where do we fail?' Why the impact of all major climate change summits is not coming at the ground level?. What more is to be done and how? are few of the questions which remain unanswered and looking for the expertise of may be psychologists or behavioural scientists to come forward.

> If we do not succeed in putting our message of urgency through today's parents and decision makers, we risk undermining our children's fundamental right to a healthy, life-enhancing environment. Unless we are able to translate our words into a language that can reach the mind and hearts of people young and old, we shall not be able to undertake the extensive social changes needed to correct the course of development. (Brundtland, 1987: xiv)

While considerations of the social, ethical, and cultural contexts for evaluating the long-term risk of climate change are emerging, relatively little attention has been paid to the psychological aspects (Dessai et al., 2004; Lorenzoni et al., 2005). Even two decades after of Brundtland's statement to look for synergic efforts from social and behavioural scientists for managing the changing climate, there is still a need to reach in people's minds through their hearts, or by managing their emotions (Midden and Wilke, 2005).

The reduction of the emission of greenhouse gases requires changes in society. Societal changes go hand in hand with public support, and here that means accepting the seriousness of the situation, government intervention, outlines for policy, and the willingness to change behaviour. Information intended for consumers and other target groups plays an important role in this.

A workshop was convened on 12 November, 2004 at Princeton University with a view towards addressing the psychological dimension. The workshop was motivated by the notion that environmental policies that target human behaviour should incorporate insights about behavioural change and decision-making, i.e., topics central in behavioural sciences. The objective of the workshop was to inform climate scientists about the current state of understanding of how people think about global warming, and hopefully, initiate a collaborative dialogue between scientists climate and behavioural.

Article 6 of UNFCCC also stresses for a country driven, cost effective, holistic and phased approach following an interdisciplinary perspective. As a part of the amended New Delhi Work Programme (2007) to implement Article 6 of the convention, six elements starting from education to training, public awareness, public access to information, public participation and international cooperation were emphasized. Under public awareness, a special effort to foster psychological/behavioural changes has been stressed (Draft decision, 13/2007).

Figure 3.2 A Typology of People's Attitudes to Climate Change

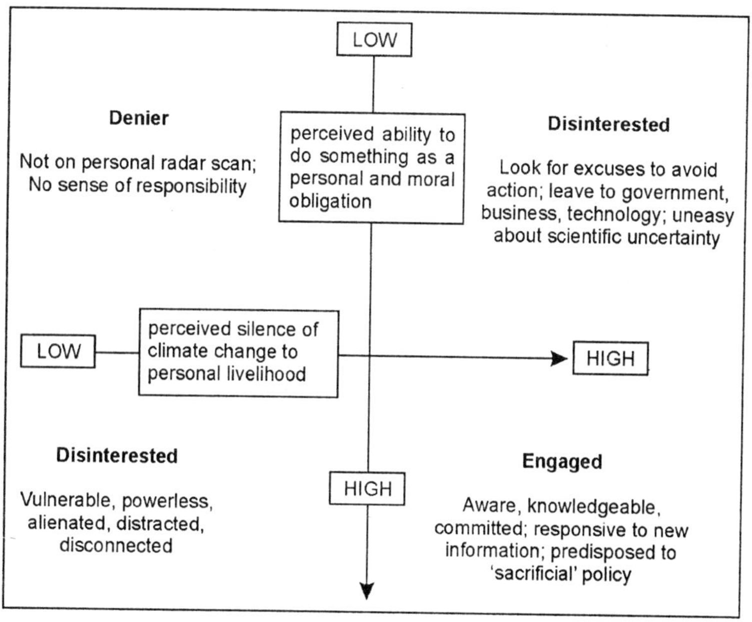

Derived from Questionnaires and Focus Groups in the Norwich Area, UK.

Source: Irene Lorenzoni, University of East Anglia, unpublished

Until recently, mainly biological technical and economic aspects of environmental problems have been illuminated. However, psychological aspects are now being widely recognized. When analyzed from the psychological perspective, environmental problems can be conceived of as caused by behavioural choices (Midden, 1993). On the other hand it is obvious that individual citizens within a given nation sometimes express deeply differing levels of awareness of and concern about climate change when asked to express their views. For example, individual citizens polarize into a typology of 'deniers', 'disinterested', 'engagers' and 'disengagers' (Figure 3.2).

This whole analysis clearly spells out the need for psychological interventions in addressing the problem of climate change. Since climate change is going to seriously affect human beings directly or indirectly in the future, it is very necessary to explore what humans think about it, i.e., their attitudes and perceptions about risk. It can be understood firstly through the analysis of human psyche and how seriously humans concerned about the possible impacts on generating to come, and their emotional involvement with the problem. In short, the affective concerns of people. Secondly, the cognitive aspects of human psyche are also very important. If people are not properly aware of the nature of the problem of climate change and how it is going to affect them, they will never be emotionally concerned about the whole issue. After cognitive understanding and affective concerns, the need for actual action or conative component arises, to resolve the problem. Little acts to save the planet add up to give big returns. In view of the above, the amended New Delhi work programme, the implementation section indicates the need to conduct knowledge-attitude-behaviour surveys to establish a baseline of public awareness which can serve as a basis for further work and support monitoring of the impact of activities.

SOCIAL DILEMMA OF CLIMATE CHANGE

Social dilemma is a group situation when personal interests are at odds with collective interests (Van Lange et al., 1992). Each member of the group is tempted to act out in self-interest to maximize personal outcomes. However, each self-interest choice creates negative outcomes for group as s whole. When a large

Figure 3.3 Behavioural Traps to Climate Change

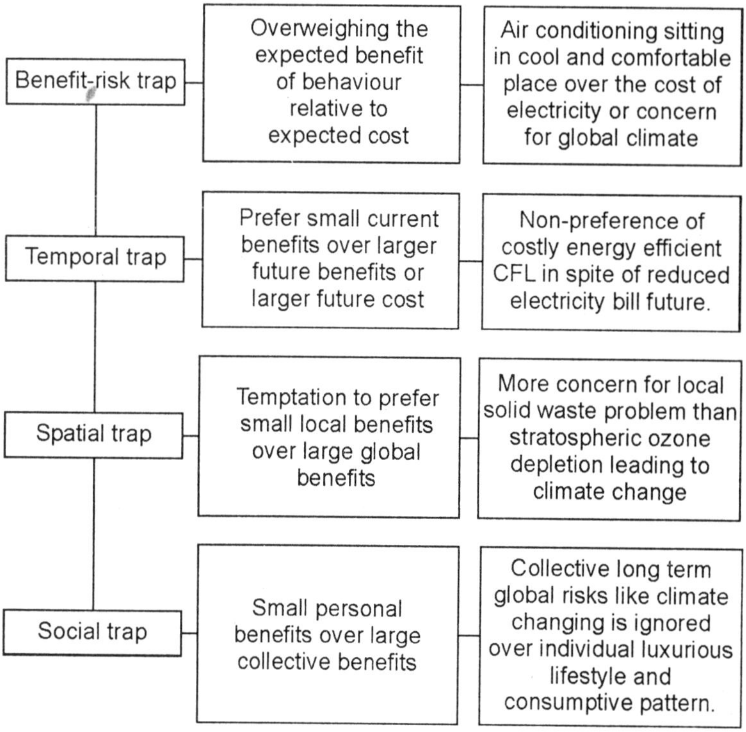

number of people make a selfish choice, negative outcomes accumulate resulting in multiple traps. Pro-climate behavioural choices are obstructed by these traps as individual's immediate temptations overweigh the expected benefit for a sustainable planet (Vlek and Keren, 1992). Figure 3.3 describes these behavioural traps in regard to climate change.

BEHAVIOURAL DILEMMA OF DEVELOPING NATIONS

In view of the common dysfunctional approach towards climate change and its impact, which is primarily guided by flight reactions, the question arises as to how can we motivate people to develop a cognitive understanding, affective concern and conative or action based approach towards climate change. In regard to developing countries, Maslow's (1943) need hierarchy has great relevance. The decade old tussle between developing and developed countries regarding reducing GHG emissions can also be interpreted through need hierarchy. Developing countries claim that when their per capita emissions are much less than developed countries, and that they are at the lower ends of developmental hierarchy in terms of education, infrastructure and per capita resource availability, it is very difficult to create the cognitive, affective and conative level concern of people towards climate change and its impacts, besides bearing the additional cost of reducing emissions and generating resources for adaptation.

Developing countries strongly feel that until and unless they are able to fulfil the basic developmental needs of their country, they cannot think about bearing any additional cost for mitigating the effects of climate change. In this regard, Indian Science and Technology Minister, Mr Kapil Sibbal remarked, at the recent India-UK CEOs' conference on climate change, that the real challenge is to provide green technologies for the small scale sector while keeping them competitive. Bigger industries will opt for cleaner technologies to improve efficiencies, but small enterprises would not if they are not cost effective. He further commented that India's first priority is to eradicate poverty and not mitigate the problems posed by climate change (*HT*, 21st January, 2008). However, looking at the huge concentration of population in the developing regions, in spite of the lower per capita emissions, the impact that they can create in the global climate is quite substantial. In view of that, the remark of British Minister for Trade, Digby Jones is also worth to be mentioned. "If the goal of eradicating poverty is not to be tampered with and while doing that, the result is huge damage to the planet, then for whom would poverty be eradicated?" He further stressed the need for the balanced approach to development and environmental conservation for which he recommended a joint effort of developing and developed countries. The stance of developing countries for targeting the basic developmental needs of their population before moving at higher levels of climate concern can not be totally ignored, and as a result, an adaptation fund has been created in recent CoP–13 (2007) to address this problem. However, looking at the steadily approaching threat to global climate which is likely to affect the developing countries in a much adverse way. And looking at their limited resources, there is a need to simultaneously address different developmental needs of the hierarchy by paying proportionate attention to restrict the emission levels too.

HUMAN PSYCHOLOGY OF CLIMATE CHANGE

In spite of the best possible efforts around the globe to combat climate change, we feel that people are still not as seriously alarmed of the expected future risk as they should be. This phenomenon varies substantially across cultures and can be explained through principles of human psychology.

Motivational Issues

It has long been assumed that an individual's behaviour in a social dilemma situation is guided by the desire to profit as much as possible from a collective good. However, according to Wilke, (1990) greed is not the only motive underlying an individual's decisions in a social dilemma. Two more motives are important as well: the equity motive and the efficiency motive. The equity motive refers to the desire that profits are distributed in a fair way among group members. The efficiency motive implies that individuals strive for a continued existence of a collective good. These latter two motives are assumed to restrain the tendency to maximize individual profits.

The efficiency motive implies that if individuals believe that the continued existence of a collective good is seriously threatened, they will decide to preserve it. Hence, if individuals believe that climate change seriously threatens conditions of life, they will be more likely to support the mitigation policy. It is, therefore, important to gain insight into how individuals appraise climate change and how they react toward it as far as mitigation strategies are concerned.

PSYCHOLOGICAL INERTIA

Another behavioural dimension may be the psychological inertia, which means that humans are quite resistant to alter their pre-existing ways of consumption and lifestyles unless a very as strong force is exerted against them threatening their mind set and psyche. Attempts for promoting behavioural modifications to check anthropogenic causes of climate change do not succeed at the rate they should be because of psychological inertia. The "Why should I change?" attitude is the outcome of either lack of cognitive understanding of the magnitude of problem or the non-assimilation of factual information about future risk from climate change in human mind due to psychological denial. It is common for people to experience various emotions and psychological symptoms when they are faced with information about climate change related threats and predictions of abrupt climate change. People may feel anxious, scared, sad, depressed, helpless, hopeless, frustrated or so. Sometimes people may get along by denying the problem or avoiding to think over the problem.

Role of Spiritual Coping

Psychological denial varies across cultures. The cultures with a more collaborative and spirituality based approach to life, are more prove to psychological denial as they have a very safe target to attribute all natural happenings over to the Almighty. Western world will definitely attribute spiritual coping as an alternative to factual knowledge which may be partially true as well. However, spiritual coping strategy to climate change has positive outcome too, in the form of more environmental resilience and increased capacity to adapt with untoward natural happenings. The inner peace and the tendency to attribute all happenings to God's wishes and reciprocating it by spending more time in religious prayers gives them strength to cope with the negativity of natural environment in a more effective way. More so, demonstrative data based factual information about changing climate is extremely important to counter the psychological denial.

Role of Locus of Control

Cultural variations in attributional style also play an important role in this regard. People or cultures with external locus of control will attribute possible risk from changing climate to God or other external factors beyond their control and thereby protect themselves from associated negative emotions. On the other hand, people or cultures with internal locus of control will attribute climate change to the modifiable societal/individual behavioural patterns. It is definitely a responsible and desirable way of behaving but the possibility of perceiving higher risk and associated negative emotions is also high which may be individually undesirable but socially desirable.

In order to understand how individuals come to see a certain situation as a threat, we may now turn to risk perception followed by the stress theory.

RISK PERCEPTION

In regard to climate change, risk perception on the part of the likelihood of public is the likelihood of perceiving the possible risk associated with the changing climate. The perception of risk is highest at the micro-level when a person is threatened with some specific impact of changing climate on him or his immediate relationship circle. It gradually reduces its strength as it extends its circumference from person to town to state to country to the globe. Meertons et al. (1994) say that people perceive high risk only when they foresee a potential degree of harm or fatality, physical extent of damage; social extent of damage, ambiguity of undesirable consequences, uncontrollability of consequences, experience or familiarity with the possible consequences etc. It means that the more we are cognitively aware about climate change and the possible adverse consequences, the more risk we shall be perceiving. That is the reason why scientists and climatologists are much more worried about the future of the planet due to climate change, than the general public. Although the study of risk perception considered primarily health and safety risks, psychometric method has been applied to characterize ecological risks as well (Mc Daniels et al., 1997).

A wide range of qualitative and quantitative methodologies have been applied to map people's perception about climate change. A considerable part of the public believes that climate change has already occurred (Weber, 1997). They try to relate it with weather related fluctuations which they experience and perceive that a few degree of mean temperature rise may not cause any adversity to planet. They also perceive them as just statistical speculations without realizing the large-scale possible geophysical and ecosystem effects (Bostrom et al., 1994).

Public risk perceptions can fundamentally compel or constrain political, economic, and social action to address particular risks. Public support or opposition to climate policies (e.g., treaties, regulations, taxes, subsidies) will be greatly influenced by public perceptions of the risks and dangers posed by global climate change. Leiserowitz (2006) conducted a national, representative survey of the US public and found that Americans have moderate climate change risk perceptions, strongly support a variety of national and international policies to mitigate climate change, and strongly oppose several carbon tax proposals. Drawing on the theoretical distinction between analytic and experiential decision-making, this study found that American risk perceptions and policy support are strongly influenced by experiential factors, including affect, imagery, and values, and demonstrates that public responses to climate change are influenced by both psychological and socio-cultural factors.

Now the issue is, whether it is appropriate to create a fear psychosis about climate change and use it as a strategy to promote micro-level pro-climate behaviour. Genuine data based public information must be provided to public which may create a state of arousal. In this regard, drive reduction model of Higbee is worth to be cited. He states that fear appeal is a potent strategy to change the attitudes which is being used by the media to promote awareness and concern about certain issues like alcoholism, smoking or so which public otherwise ignores.

CLIMATE CHANGE AS A STRESSOR

Appraising the possible environmental events as harmful, threatening or risky may give rise to physiological, emotional, cognitive and behavioural changes (Taylor, 1986). According to Lazarus and Folkman (1984), people engage in primary as well as secondary appraisal processes which eventually result in the subjective experience of stress. Primary appraisal with regard to climate change induced stress will be the likely effects of climate change as negative, harmful and stressful at present or in future. The secondary appraisal process in climate change is linked with the available coping resources and adaptation process. People or societies capable of adapting with possible effects of changing climate may perceive low stress as compared to societies with limited resources for adaptation. However, alternatively, such societies may also develop learned helplessness and resort to exclusively spiritual coping if they are passive. On the other hand, active and well-aware societies will try to develop adaptation mechanisms of their own for future or negotiate with other potential countries to develop a common pool of resources to combat climate induced stress. The recent 'Adaptation Fund' of UNFCCC in CoP–13 at Bali is the outcome of such efforts of active and aware countries of the developing world.

The whole phenomenon of climate change-induced stress is built upon the behavioural principle that "stress is indispensible". There will be no active effort on the part of world to combat climate change if there is no negative perception associated with it. Once, there is stress the whole world gears up to combat it. Moderate amount of stress increases positive efforts to overcome it as per Yerks Dodson Law. The same is happening around the world, though coordinated efforts to empirically explore these behavioural dimensions are needed for which psychologists should take the lead.

CONCLUSION

The thorough analysis of various behavioural dimensions starting from psychological inertia to risk perception, psychological denial, fear psychosis and associated primary and secondary appraisal of stress indicates that climate change is deeply associated with human psyche. Human response to changing climate and coping/adaptation efforts at micro and macro levels will thoroughly depend on how we are able to positively manage the human psyche for the benefit of future societies. Human psyche, if positively charged, has the potential to manage anthropogenic causes of climate change in a way that is needed for a sustainable planet, the earth. What is really needed is that people should have the cognitive understanding of the climate scenario, are able to hear their inner voice to relook at their behavioural choices. This is what the earth is looking forward to.

REFERENCES

Blakemore, B. (2005) "The Psychology of Global Warming: Alarm-ist Versus Alarm-ing". ABC News Thursday 8th Dec.2005. http://www.trusthout.org/cgi-bin/artman/exec/view.cgi/34/16113 cited on 15.01.2008.

Bostrom, A., Morgan, M.G., Fischhoff, B., and Read, D. (1994) "What Do People Know About Global Climate Change? 1 Mental Models". *Risk Analysis*, 14, 959–970.

Brundtland Commission Report (WCED) (1987) Oxford. Oxford University Press.

Christensen, J.H., Hewitson B., Busuioc A., Chen A., Gao X., Held I., Jones R., Kolli R.K., Kwon W.T., Laprise R., Magaña Rueda V., Mearns L., Menéndez C.G., Räisänen J., Rinke A., Sarr A. and Whetton P. (2007) "Regional Climate Projections". In: *Climate Change 2007: The Physical Science Basis.* Contribution of Working Group I to the Fourth Assessment Report of the Intergovernmental Panel on Climate Change [Solomon S., Qin D., Manning M., Chen Z., Marquis M., Averyt K.B., Tignor M. and Miller H.L. (eds.)]. Cambridge University Press. Cambridge, and New York.

Cruz, R.V., Harasawa, H., Lal, M., Wu, S., Anokhin, Y., Punsalmaa, B., Honda, Y., Jafari, M., Li, C. and Huu Ninh N. 2007, *Asia. Climate Change Impacts, Adaptation and Vulnerability.* Contribution of Working Group II to the Fourth Assessment Report of the Intergovernmental Panel on Climate Change, Parry M.L., Canziani O.F., Palutikof J.P., van der Linden P.J. and Hanson C.E. (eds.), Cambridge University Press. Cambridge, pp. 469–506.

Dessai, S. Adger, W.N., Humle, M., Turnpenny, J. Kohler, J., and Warren, R. (2004) "Defining and Experiencing Dangerous Climate Change", *Climatic, Change* 64, 11–25.

Fransen, W. and Janssen, L. (1998) "Greenhouse Policy is CO_2 Policy". *Climatic Change*, pp. 14–16.

Harris, F. (2004) *Global Environmental Issues*. England. Wiley.

IPCC (2007) IPCC. 2007, Fourth Assessment Report, Intergovernmental Panel on Climate Change Secretariat. Geneva, Switzerland. http://www.ipcc.ch/

Lazarus, R.S., and Folkman, S. (1984) *Stress Appraisal and Coping* New York, Springer.

Leiserowitz (2006) "Climate Change Risk Perception and Policy Preferences: The Role of Affect, Imagery, and Values" Climatic Change. Vol. 77, Numbers 1–2, pp. 45–72.

Lorenzoni, I., Pidgeon, N.F. and O'Connor, R.E., 2005, "Dangerous Climate Change: The Role for Risk Research". *Risk Analysis* 25, 1387–1398, DOI. 10.1111/j1539–6924.2005.00686.x.

Lorenzoni, Irene (2004) A Typology of People's Attitudes to Climate Change Derived from Questionnaires and Focus Groups in the Norwich Area, UK. Unpublished dissertation, University of East Anglia.

Martens, P. Kovats, R.S., Nijhof, S. p.de Vries, Livermore, M.T.J., Bradeley, D.J., Cocks, J. Mcmichael, A.J. (1999) "Climate Change and Future Population at Risk of Malaria". *Global Environmental Change*. Vol. 9, No.1, 89–107.

Maslow, Abrahan H. (1943) "A Theory of Human Motivation". *Psychological Review,* 50 (4), 370–96.

McDaniels, T.L., Axelrod, L.J., Cavanagh, N.S., and Slovic, P. (1997) "Perception of Ecological Risk to Water Environments". *Risk Analysis*, 17, 341–352.

Meertens, R.W., Pligt, J. Van Der, and Vlek, C.A.J. (1994) Omgaan met milieurisico's: Beoordeling, communicatie en besluitvorming (Coping with environmental risks: "Judgment, Communication, and Decision Making"), In C.J.H. Midden and G.C. Bartels (eds.), *Consument en milieu: Beoordeling van Milieurisico's en Sturing van Gedrag "Consumer behaviour and Environment: Perception of Environmental Risks and the Influencing of Behaviour",* pp. 57–80, Houten/ Zaventem, The Netherlands, Bohn Stafleu van Loghum.

Midden and Wilke (2005) Climate Change and Changing Attitudes Effect of Negative Emotions on Information Processing, Unpublished dissertation, Eindhoven University of Technology, Faculty of Technology Management.

Midden, C.J.H. (1993) Milieubedreiging als gedragsmotief: Cognitieve en sociale oorzaken vanonderwaardering "Environmental threat as behavioural motive: Cognitive and social causes of underestimation". *Milieu*, 1, 8–13.

Pachauri, R.K. (2004) "Climate and Humanity". *Global Environmental Change*, 14, pp. 101–103.

Sheppard (2003) "Sea Surface Temperature 1871–2099 in 14 Cells around the United Kingdom". *Marine Pollution Bulletin,* Volume 49, Issues 1–2, pp. 12–16.

Sibbal, Kapil (2008) Editorial in *The Hindustan Times* New Delhi Edition, 21st January, 2008.

Stott, S. and Sullivan, S. (2000) *Political Ecology: Science, Myth and Power*. London, Arnold.

Taylor, S.E. (1986) *Health Psychology*. New York, NY, Random House.

UNFCCC, 2007 UNFCCC. (2007) "Vulnerability and Adaptation to Climate Change in Small Island Developing States": Background paper for the expert meeting on adaptation for small island developing states. UNFCCC Secretariat. Bonn, Germany.

http://unfccc.int/files/adaptation/adverse_effects_and_response_measures_art_48/appli- cation/pdf/200702_sids_adaptation_bg.pdf

Van Lange, P.A.M. and Messick, D.M. (1996) "Psychological Process Underling Cooperation in Social Dilemmas". In Gasparskhin, M.K., Mlicki and B.H. Behethy. (eds.) *Social Agency: Dilemmas and Education Praxiology*, Vol. 4, Transaction Publisher, pp. 93–112.

Vlek, C.A.J., and Keren, G.B. (1992) "Behavioural Decision Theory and Environmental Risk Management: Assessment and Resolution of Four Survival Dilemmas". *Acta Psychologica*, 80, pp. 249–78.

Wassmann R, Nguyen Xuan Hien, Chu Thai Hoanh and To Phuc Tuong (2004) "Sea Level Rise Affecting the Vietnamese Mekong Delta: Water Elevation in the Flood Season and Implications for Rice Production". *Climatic Change*, 66 (1–2), pp. 89–107.

Weber, E.U. (1997) "Perception and Expectation of Climate Change. Precondition for Economic and Technological Adaptation". In M.H. Bazerman, D.M. Messick, A.E. Tenbrunsel, and K.A. Wade-Benzoni (eds.) *Environment, Ethics, and Behaviour. The Psychology of Environmental Valuation and Degradation.* pp. 314–41, San Francisco, CA, The New Lexington Press.

Wilke, H.A.M. (1990) *Afstemming Van Individuele en Groepsbelangen (Tuning Individual and Group Interests),* Rede uitgesproken op 9 November 1990 bij de aanvaarding van het ambt van gewoon hoogleraar in de sociale psychologie aan de Rijksuniversiteit Leiden (inaugural lecture), Leiden, The Netherlands, Rijksuniversiteit Leiden.

Wilkinson C. (2004) "Status of Coral Reefs of the World: 2004," V. 1. *Australian Institute of Marine Science*. Townsville, Australia. 302 pp.

Yechuri, S.R (2008) "Mercury Rising". *The Hindustan Times*, Thursday, January 24th, 2008.

4

Coping with Urban Environmental Stress

GAYATRI KHUNTIA AND PARUL RISHI

Big urbanized cities are characterized by a plethora of excitement, too much noise and movements, air loaded with pollutants and dust, lands loosing its natural beauty, high skyscrapers competing with each other obstructing the view of beautiful sky and too much density where every day the city is facing high cost of living, astronomical amount of rents, persistent housing crisis, pollution, noise, anxiety, nervous fatigue and incredible consumption of tranquilizers and sleeping pills. The soul and body of the city is scarified to the so-called necessities of automobile traffic.

The central issue of the urban challenge and urban stress is familiar to every city dweller where every one encounters the physical environment not as a satisfying and enriching resource, but as a source of psychological stress posing the challenge of coping effectively with extreme environmental demands. What is at risk are psychological consequence of the failure to master the environmental stress and eventually the hidden psychic costs of long-term adaptation to adverse environmental conditions.

Urban population, especially economically weaker sections, suffer from various types of environmental stressors and there is an urgency to guide urban development towards sustainability to promote quality of life and well-being among city dwellers.

Environmental problems and urban deterioration are the frequent results of shortcomings in the planning and urban management process, institutional frameworks and governance and the inadequate allocation of financial resources in addressing the environment concern. This dual trend of rapid population and economic growth is responsible for massive demographic shift from rural to urban societies, raising enormous issues in terms of sustainability, efficiency and equity. It is measured in terms of Environment Sustainability Index (ESI) measuring overall progress towards environment sustainability through various indicators. ESI mainly focuses on five components and environment stress is one of them (de Sherbinin, 2003). The globalization regime has lead to an increase in income in the cities, however, at the same time, an increase in poverty, inequalities, regional disparities and, above all, environment stress of various types. In this chapter, the nature and dynamics of urban environmental stress and coping potential of people will be analyzed.

URBAN CITIES: DESTINATION FOR WHAT?

"Almost everywhere, cities are the destinations for people escaping poverty, conflict and human rights violations, or simply those looking for ways to build better lives," said the UN Secretary-General Kofi Annan, on World Habitat Day, 2006 In the past, cities have been the pinnacles of civilization and the hubs organizing and exploiting their hinterlands. Their bricks and mortar, and streets and spaces are the result of never-ending competition and change. In the future, cities are quite likely to change in ways that are hard to imagine. The physical environmental settings that encompass our daily life, reflect very personal and meaningful aspects of us and they have remarkable influence on the fabric of our life. As we enter the new millennium, developing countries face a number of burning issues and challenges. Urbanization entails major demographic, social, cultural and environmental transitions and changes that have a marked impact on the health of the populations concerned. Great pressure on arable land is leading to unprecedented population movements, including an explosive growth of urban centres. Throughout the world, urbanized areas are expanding into surrounding areas which are under agricultural or various natural land covers such as forests, wetlands or grasslands. These changes may have significant impacts on the ecosystem services, biodiversity (through habitat loss and landscape fragmentation), hydrological systems, and local climate which, in turn, may impact human health. Furthermore, increasing reliance on automobile-based modes of transport leads to emissions of smoke and dust, problems of congestion in traffic, problem in parking ground etc.

THE CONCEPT OF ENVIRONMENTAL STRESS

Stressors are aversive, primarily uncontrollable and of variable duration and periodicity and require low to moderate adjustment. These refer to any force or event in the human or natural environment that may cause a person to experience stress.

Some examples of environmental stressors include: loss of trees, land degradation, declining water quality, mineral depletion, noise, air pollution, crowding, traffic congestion, terrorism, natural disasters, extremes of temperature and land set apart for decomposing of waste products. Studies on the effect of different environmental stressors on people indicate that they can impact people's behaviour, mood, cognitive function, physical health and/or psychological well-being. It is the condition interfering with optimal human functioning and is appraised as threatening for today or for the future.

Behaviour is a function of persons, the environment and the interaction between the two. Therefore, if a person's environment is not congenial towards him, his/her behaviour is reflected in the form of stress. The stress provided by the urban environment often reflected in the form of physical turmoil, psychological illness and reduced social interaction.

Research in environmental stress offers an essential aid in efforts to humanize the urban environment and more effectively to meet the needs of the people. Looking at the problems in this area and their long-term consequences, the urban environmental stress is a uniquely important field of study in relation to subjective well-being. Ironically, urbanization and rapid population expansion were themselves catalyzed by a profoundly anti-urban vision for the region. Degradation of the ambient environment has also been severe, as a result of industrialization, vehicular traffic, and overconsumption of energy and water. Though urbanization is, to some extent, responsible for improving the status of the people, simultaneously it has lead to poor physical and mental health too. So, it is a trade-off which is to be patiently evaluated and judged.

THE ROLE OF URBANIZATION

The amount of stimulation provided in the cities in the form of noise, crowding, congestion, pollution impinge on the arousal and adaptation levels of individuals and may lead to the negative consequences associated with exceeding optimum levels. Barker's (1968) behaviour setting approach indicates that cities present problems with providing meaningful roles for all inhabitants, thus marginalizing and alienating a good many people. In addition, cities are associated with increased isolation and smaller, less close social support network. There is a strong evidence that city dwellers learn to screen information in order to cope with the hustle and bustle of city life. Urbanization has an effect on social behaviour too. City life involves a very complex and dynamic series of consequences, and these are very difficult to quantify. Globalization has led to rapid industrialization and urbanization in developing countries, such as India, and an increase in lifestyle-related disorders, such as coronary heart disease, essential hypertension, stroke, and diabetes mellitus, in these countries (Gupta, 2004; Reddy, 1999). Stress caused chronic ailments encountered by people residing in the neighbourhoods increases vulnerability to immune and cardiovascular disorders. (Gump et al., 1999).

Environmental stress is a relational concept signifying an imbalance between environmental opportunities. Flood, earthquakes, volcanic eruption, major storms, nuclear power plant accidents, and discovery of toxic waste dumps are examples of cataclysmic events that demand major adaptive response from all individuals directly affected by the event (Baum and Singer, 1982; Campbell, 1983; Lazarous and Cohen, 1977). The mediating role of expectancies is dramatically illustrated by research on crowding and noise, showing that anticipation of these environmental stressors cause reactive symptoms and striking similarity to actual exposure to stressors (Baum and

Paulus, 1987). These anticipated stressors' effect include performance aftereffects, physiological arousal, coping behaviours, and negative effects. Chronic conditions that environmental stressors can cause inner conflicts and stressors that can lead to existential crises challenging the meaning in life (Wong, 1993).

URBAN ENVIRONMENTAL STRESS

Urban environment stress is the typical situation that represents the level of anxiety of the people who are residing in the urban environment and facing the daily conundrum that is incompatible with their life. Life in the urban settings is an interphase between the adverse physical settings as described earlier and the individual characteristic which arise from the time spent in the exposure to such aversive situations like time spent in commuting, time spend in crowd, exposure to polluted air and smokes, exposure to high level of noise etc. If these situations fall within the tolerance level of an individual then the person does not experience stress, but if it exceeds the acceptance level then the person is said to be under stress. A recent study on urban stress in the slums of New Delhi, has revealed that individual-level stressors assessed via crowding in the home affected women more than men, whereas men rated neighbourhood-level stressors such as traffic, noise, and pollution, as being more stressful than did the women (Ruback and Pandey, 2002). In that study, individual characteristics, such as age and psychological variables, are taken into account, and there are significantly affected reports of both mental and physical distress.

The City as a Stressor

The experience of living in a city involves an increase in all of the major stressors. Cities are the greater source of noise leading to greater level of annoyance, irritation and insomnia. Like noise pollution is associated with negative effect and interpersonal behaviour. The congested habitat in the city, the encroachment by buildings and traffic, all symbolizes crowding. Crowding is associated with aggressive behaviour, anxiety, tensions and stress (Baum and Paulus, 1987; Sundstorm, 1978). However, household and neighbourhood characteristics of the affected report mental distress rather than physical symptoms. Community-level contextual stressors, such as poverty, unemployment, crime, and lack of social support, have been associated with both an increased incidence of behavioural and emotional problems and inadequate mental health care (DosReis and Harmen, 2001). Several studies have examined the impact of community-level contextual stressors on behavioural and emotional problems of children and adolescents (Aneshensel and Sucoff, 1996; McLeod and Edwards, 1995; Shumow et al., 1998). Researchers (Deardorff, Gonzales, and Sandler, 2003; Safyer, 1994) have posited that life in the inner city is characterized by numerous factors (e.g., violence, poor housing, drug use) that can affect the development of the adolescents who are exposed to those factors on a daily basis.

Cities provide continuous more intense and more complex sources of stress which challenge the coping resource of individuals owing to isolated, marginalized and anxious lifestyle (Halpern, 1995; Veitch and Arrkelin, 1995). Big cities are associated with increased rate of isolation, alienation, limited social support networks and surprisingly greater incidences of psychological disorders especially in poor areas (Cochrane, 1983). As such, it is not surprising that cities have been credited with greater incidence and prevalence of mental disorders (Crowell, 1986). Within the city high rate of mental disorders tends to be concentrated in the poorer areas of the inner city (Dean and James, 1984; Levy and Rowitz, 1973). However this inner-city effect is probably better explained in terms of social drift (Cochrane, 1983). City dwellers are also less likely to help someone in distress, which may be more to do with the presence of others than city living per se (Korte and Kerr, 1975; Lantane and Darley, 1968). Many people in recent years are living under conditions involving various kinds of major and minor stress.

Studies on crowding (Paulus, 1980; Epstain, 1982) noise (Cohen and Weinstein, 1982), air pollution (Evans and Jacob, 1982) and heat (Bell and Greene, 1982) have found a complete or partial amelioration of many negative impacts of exposures to environmental stressors. An extensive body of literature deals with noise as an urban environmental stressor (Cohen, Glass and Philips, 1979; Glass and Singer, 1972 and Jenkins, 1993;

Campbell and Schwarz, 1996; Guerra et al., 1995). Such incidents can keep children confined to their homes. Witnessing an act of violence can lead to post-raumatic stress disorder (Bell, 1987; Garbarino, Kostelny, and Dubrow, 1991; Walker, Goodwin, and Warren, 1992), and violence can affect children in multiple and prolonged ways. Findings suggest that urban reforms generate economic gains at the expense of urban residents' psychological well-being. Ostfield and D'Atri, (1977) have investigated urban stress and concluded that a higher crime rate was associated with high population density.

COPING WITH ENVIRONMENTAL STRESS

The negative aspects of life have drawn the attention of health and human service professionals for a long time. They have focused on reducing urban flight, crime, pollution, and traffic congestion. It became increasingly obvious that communities could not flourish by focusing only on their needs and deficiencies (Kretzmann and McKnight, 1993). The shift away from an exclusive focus on deficits, a focus on the well-being of the community inhabitants is also essential. Therefore, different ways are adopted to improve the health promotion and well-being. The way adopted to manage an event or stimulus perceived as threatening is part of coping.

Coping is an important aspect of dealing with stress. How a person sees or appraises the situation will determine how they will manage the situation and the resulting attitude. A person needs to know where stress is coming from and what it is (Wong, 1993) so that effective coping strategy can be employed. Development of coping abilities that can make a person's personal adaptation successful is of importance as it can influence his or her self-confidence and outlook in life.

Coping could minimize the harmful effects of the stressful situation and could enable one in meeting the demands of everyday life and fulfilling meaningful life-goals (Wong, 1998), facilitate health and energy (Lazarus and Folkman, 1984), and it is essential for the emotional process and the emotional life (Lazarus, 1999). Advocating the importance of positive events including leisure as a means of coping with stress, Folkman and

Moskowitz (2000) has argued, "Historically, coping has most often been evaluated in relation to its effectiveness in regulating distress". The experience of positive events (e.g., getting together with friends) also potentially helps to generate positive meaning in the midst of stress (Folkman and Moskowitz, 2000). Theory of coping explores out some coping mechanisms to deal with issues such as overcrowding and conflicts in selecting outdoor recreational activities (Manning et al., 2001). People living in unnatural and stressful conditions need yoga to combat the stressful life they face on a daily basis (Nespor, 1987). Impersonality and aloofness are one of the ways by which people cope with overcrowding. This can bring about unneighbourliness and may lead to increasing remoteness, loneliness and stress. Both internal and external awareness is advocated by yoga, (Swami Satyananda, 1980) and is also helpful when coping with stress.

Some studies have focused on the degree of positive emotions impacting perceptions of stress and the use of coping strategies. The positive affect associated with enhanced immune function is derived from many sources, including sensory pleasures, appreciating beauty, rewarding relationships, or involvement in intrinsically motivated activities through which one experiences a sense of mastery or self-determination (Charnetski and Brennan, 2001). Leisure activities are undertaken by choosing activities like recreation, attending social gathering, and religious pieties or some physical exercise etc. There are various modes of coping that people adopt to relieve the stress. Some of the common coping techniques are mentioned below.

Social Coping

There is extensive evidence of positive social relationships contributing to positive affect and well-being. The need for regular, pleasant, ongoing social relationships is a "powerful, fundamental, and extremely pervasive human motivation" (Baumeister and Leary, 1995). This rationale is used to hypothesize that participation in leisure activities may produce friendships and contacts that lead people to perceive that they will receive social support when necessary.

Environmental Coping

The environment has a significantly great effect, particularly the woodlands being advantageous, especially for positive emotions. Science and medical research have clearly demonstrated that recreation pays when it comes to our overall health and well-being. Being close to nature relieves all the stress and gives to solace from the complexities of life.

Religious Coping

Since the Vedic period, chronic illnesses and stress has been cured by Yoga, meditation and relaxation. Spirituality facilitates intra-personal relationships, and makes man more skillful in facing life's various situations. Even rituals associated with spiritual life have often proved to be effective stress reducers. Yogic techniques have the potential to cure various stress related diseases; mindfulness meditation helps stress reduction, improving physical and mental health, improve concentration, creativity, and above all a sense of well-being and calm.

Physical Coping

Coping by physical exercise is also a way of relieving stress. Exercises and sports like climbing, swimming, aqua aerobics, gym, and volleyball leagues release negative energy to refresh people. Leisure is an important focus in therapeutic recreation services and can serve as an avenue through which people can increase their hedonic capacity and well-being.

Cultural Coping

Visiting ancient places has had great significance as it contributes to the relaxation to moods along with bringing us close to our history. Listening to music, watching movies are some of the ways through which we can cope with environmental stress as these activities take us away even for a small fraction of time from the dreariness of life and we get peace of mind and find ourselves in a state of contentment.

ENVIRONMENTAL STRESS AND COPING: A STUDY OF BHOPAL CITY

Environmental stress cannot be alienated from general experience of stress. People in occupations experiencing high work stress associated with poor coping, will be more prone to environmental stress too. A study by Khuntia and Rishi (2007) was conducted in the urban Bhopal to assess the level and effect of urban environmental stress on the subjective well-being. The major problem perceived by people was found to be unhealthy living conditions in slums. Despite being the capital city there were no proper arrangements for slum dwellers which forced them to live a unwholesome life generating stress. Suitable arrangements for the waste management were also very limited. The heaps of garbage and unwanted waste accumulate here and there spoiling the beauty of the city. Though some developmental work had taken place in Bhopal yet that was confined to new Bhopal only. The old Bhopal area was a huddle of congested houses, constricted lanes causing a major problem to the city dwellers.

Experience of stress differs across zones as urbanization has not taken place in all the zones uniformly. Well developed infrastructures, ample supply of water, parks for recreation are mostly located in the south zone, though some parks are there in the north too. The south zone is famously known as new Bhopal and is surrounded by good urban forest, enough green coverage, planned development, wide roads, and free flowing traffic which provide people with a comfortable city environment, reducing the stress level up to a great extent. It has been seen that woodland and green spaces can contribute to people's health and well-being. According to Kaplan (1995), natural environments are particularly helpful in restoring attention, as they are gently stimulating to the senses. As BHEL is located in the east zone, the people of there are mostly live in the campus of BHEL housing colony, and are having a good surrounding environment. Hence, they are not much affected by stressors like noise, water, waste etc. But the people inhabiting the west zone experience more stress because that zone is lacking in the above facilities in comparison to other zones. The study revealed no genderwise differences in the perceived stress level. The increased globalization and rapid urbanization has brought about the era of revolution when women are also getting opportunity to work. Therefore, while working and commuting, they also face stressors up to the same extent as experienced by their male counterparts.

Personality is another factor which leads to stress (Sharma et al., 1998). People with type 'A' personality show multiple behaviour patterns, time urgency, competitiveness, impatience about their growth that causes them a greater exposure to stress of all types including environmental stress (Heilbrun and Friedberg, 1988).

It has been seen that despite experiencing urban environmental stress the people of Bhopal viewed their city as not much stressful. The reason for this is, that people always perceive the situation positive or negative in the perspective of referents. If Bhopal is compared to other cities of similar size or with metros like Delhi, Mumbai etc., where the situation is more grim and challenging, people find their life as quite satisfactory which motivate them to face existing environmental stressors in a more positive way (Adams, 1963).

A person who is stressed by the city environment will analyze his/her city in comparison to other cities. If s/he finds that s/he is at the positive end, equity is maintained. If s/he is not able to do so and finds oneself in a disadvantageous position, s/he feels inequity. The outcome is three fold. One may be compensating with coping, functional or dysfunctional to deal with or avoid unhealthy aspects of city stress as possible. Another may be behaviour or conative action, to constructively involving oneself in the improvement of urban environment and motivating others to participate. If this also fails or is not possible, the last possible action to reduce stress and to restore equity may be moving away from the city and look for other options where environmental stressors may be limited. There are enough examples of people living in a farm house away from city to avoid the city hassles and get peace of mind. Similarly, there are cases where people from villages come to cities, to get the taste of high civic amenities and opportunities, in spite of the city hassles.

Human beings have a strong need for environmental mastery and sense of self-efficacy (Averill, 1973). Negative consequences associated with lack of control include negative effect, cognitive deficits, and reduced motivation to behave instrumentally when the option is available, are called 'learned helplessness' (Seligman, 1975). The growing stress in the urban environment has created a situation among the city dwellers where they feel that the current environmental condition will be an ongoing process, they do not have any control over those factors leading to learned helplessness among the people. Here the citizens perceive that they do not have control over an environmental source of stress, or they have very little ability to influence the environmental outcomes by their own behaviour making them helpless by forcing them to face those stressors.

The experience or perception of the city is represented as a joint function of the objective environmental conditions (for example, population density, temperature, pollution levels) and the individual characteristics of the person (for example, adaptation level, previous experience, and time in the city). If the perceived environment is beyond the individual's optimal range (for example, if it is over stimulating, contains too many stressors, constrains behaviour, or offers insufficient resources), stress is experienced which, in turn, elicits coping. If the attempted coping strategies are successful, adaptation and/or habituation occurs, though possibly followed by after-effects such as fatigue and reduced ability to cope with the next stressor. Positive cumulative after-effects would include a degree of learning about how to cope with the next occurrence of undesirable environmental stimulation. Finally, a positive discussion regarding perception of the environment for future events is needed as people start thinking about life in city as full of hassles and un manageable stress.

Despite the high perceived stress, it has been found that overall coping indices were low. Unsuccessful to cope makes people aroused and possible after-effects like mental discomfort, learned helplessness and performance deficit may result (Figure 4.1).

Figure 4.1 Effect of Urban Environmental Stress on Quality of Life

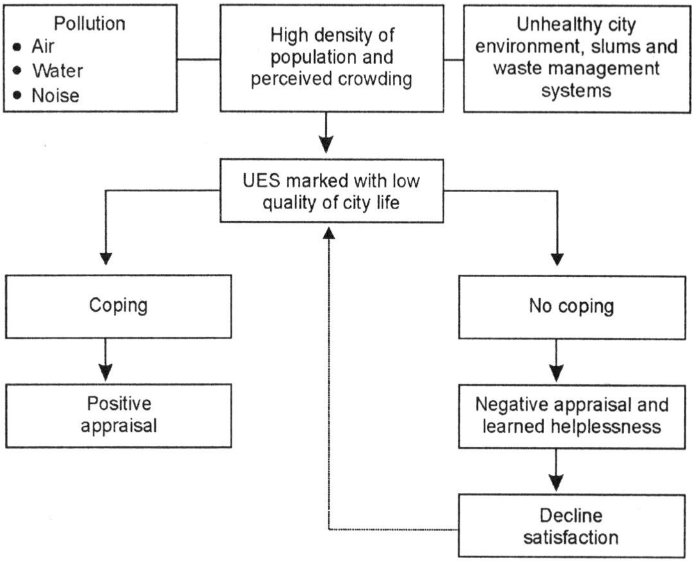

URBAN ENVIRONMENTAL STRESS

The impact of urbanization has an effect on the perception of citizens. The well planned area with ample amount of natural settings has a positive effect on the life of the city dwellers, whereas the poorly designed region deprived of natural surroundings and with all the environmentally limiting condition has lead to the consequence as UES. Individuals adopting coping successfully are able to counter with the stress. Those who are not able to cope adequately, may suffer from possible after-effects of exhaustion, learned helplessness, severe performance decrements, illness and mental disorders, which people can otherwise minimize by adopting various coping strategies.

Chronic conditions such as environmental stressors can cause inner conflicts and stressors which can further lead to existential crises concerned with the meaning in life (Wong, 1993). Chronic conditions like pollution, crowd, noise etc., act as root of the chronic mental illness and decline the satisfaction level of the people. Coping and adaptation have been related to subjective well-being. Though the people of Bhopal perceived urban environmental stress, still the subjective well-being is higher in society. The reason behind this, may be that gradual exposures to those stressors have increased the adaptation level of the citizens. Experienced stress is being managed by many of the people either by positive appraisal or comparison with the life of others urban centres.

On the other hand, for others pessimism and depression along with the negative approach towards stress bring down their satisfaction level. Evaluation of subjective well-being was done on the basis of quality of life, condition of urban living and satisfaction from surrounding environmental condition. The overall working population of Bhopal experiences the moderate subjective well-being in spite of high stress. The way of coping among different professionals was found to be least effective in the business class people because they have less available time for coping. It is in evidence that certain coping strategies, such as spirituality, positive reappraisal, optimism, and active problem solving, are linked to higher subjective well-being (Diener and Lucas, 1999; Folkman, 1997). It has been found from the analysis that people belonging to business class do not find much time for religious mode of coping which may be the reason for their lower SWB. There is a lack of positive appraisal among the business class which substantially lowers their SWB leading to least experience of well-being. Social coping was also least present signifying that they do not have much time for community interaction which is an indicator of quality of life. Their satisfaction from work and productive environment was also very low because they consider Bhopal as not suitable for business transaction as compared to other cities leading accounting for their low SWB.

There is a close association between nature and human well-being, therefore, realizing that nature and appraising stress in positive way through coping by appreciating the surrounding environmental resources can increase the quality of life of human beings. The progress of any urban civilization depends on the well-being of the people, by way of conserving nature, various pollution management interventions, avenues for recreation etc.

This should be kept in mind for the long-term well-being of a society to take a step further towards sustainable development.

ENVIRONMENTAL STRESS IN BHOPAL: WAYS AHEAD

Urbanization contributes not only to the economy of a civil society but also to the overall human well-being of the society for achieving sustainability in terms of real development. In urban areas of Bhopal there is no dearth of natural surroundings and rich socio-cultural and ecological heritage are also present there. However, the way they should be used to lessening the stress is not very strong. The study explored the dynamics of environmental stress as perceived by the working city dwellers primarily due to unplanned urbanization.

Micro-Level

At the micro-level some recommendations for lessening environmental stress and its effect on the citizens are suggested as the following:

- Living in the city is full of hassles therefore people should take stress in a positive way by changing the mode of appraisal. The media should publish such articles through which people may develop a sense of belongingness for the city with high internal locus of control. When people will feel themselves responsible for all the positive as well as negative environmental features of the city, they will have positive environmental appraisal and will be motivated to address the environment problems. Positive environmental attitude coupled with action is expected from the responsible citizens of Bhopal to conserve the city environment. As they cannot avoid the root causes of the stress, therefore they should try to reduce the effect of stress by effective coping.
- By expanding, exploring, and experiencing one's environment, one can appraise and appreciate the nature to reducing stress and increase one's SWB. People should make a relaxed response in stress evoking situations to promote calmness and control.

- There is need to strengthen the social support system further by maintaining healthy neighbourhood and close interactions with family and friends to improve one's quality of life.
- As people undertake less physical activities and are more prone to stress, therefore a change in the lifestyle is called for. Individuals should go for spiritually therapeutic and physical exercises like yoga and meditation, morning walk etc., for higher quality of life, enhanced mood, stress reduction and positive self-image. These are some of the ways that one can inculcate in its daily routine to fight against the daily hassles.
- In the hectic schedule of workload each and every individual should devote sometime for recreation and search for meaning to their life to help in attaining SWB.

Macro-Level

- At the macro-level while planning for the city, there should be provision for low cost multistoried accommodation for slums dwellers who are generally rural migrants. The type of life they live in the city slums not only affect their well-being but also contribute to the regression of economy, aesthetic, environment and social equity of the society. Besides, at the planning at the government level, increasing the civic facilities and opportunities for livelihood in slums as well as rural areas, will itself reduce migration. Some NGOs who are working for the welfare of the society can come forward in this regard.
- The economy of any city depends upon the number of industries in the city. While planning for industries due attention should be given for enough plantation area with natural covering and recreation sites to buffer the effect of pollution and stress.
- Any industry, while it starts functioning, should assess its impacts on the nearby environment and ecosystem. Therefore it should have the prerequisite of having a waste treatment plant if its byproducts cause pollution. The State Pollution Control Board should more strictly ensure the evasion of rules and responsibilities by the industries.

- In respect of the urban zones, with high level of air and noise pollution, force people to accept such stimulation, a proper attention should be paid by the government through stringent laws and their enforcement to minimize the noise pollution. Roadside plantation can act as barrier to obstruct the air and noise pollution. Plantation of trees species like Ashoka (*Saraca indica*), Neem (*Azadirachta Indica*), Mahogany (*Swietenia mahagoni*), Caesalpinia (*Caesalpinia coriaria),* Arjun (*Terminalia Arjuna*), Mango (*Mangifera indica*) etc., would go a long way in that direction.
- Besides, government should impose taxes and fines for more strong enforcement of laws towards reducing spitting, use of polythene, littering the road, polluting lakes, parks or recreational sites.
- There should be hoardings at various recreation sites indicating the sensitivity of that area and threat to that area due to littering with disposables, to generate awareness among the public. This sort of awareness can also be generated at schools to have widespread effects.
- The waste collected by the municipalities should be done separately for degradable and non-degradable wastes to ensure a separate treatment facility. The public should be made aware by the media for the use of separate dustbins for segregating the waste at home as well as at the public places. As the city is having very limited number of landfill sites, so allocating more space outside the city for safe disposal of waste can help eliminate the problem of waste accumulation.
- Beside these, the media should help in focusing on the issues of UES and publish articles on the UES and provide suggestions towards the role of coping to combat those stressors.

CONCLUSION

The core dilemma of urbanization has its impact not only on environment but also on the physical and mental health of human beings. The urbanized societies have to bear a lot of stressors at the cost of peace and harmony. In the light of the failure of most urban theories to include concerns for urban ecology and nature, it is not surprising that most normative approaches to urban analyses ignore nature, emphasize instead the importance of environmental justice in cities. With the growth of any city the presence of potential stressors in city's environment like noise, air pollution, crowd, waste, slums also increases. These stressors shower city dwellers with irritation, anxiety, annoyance and several others mental disorders.

The urban agenda designed to solve urban ills like poverty, traffic congestion, stress etc., includes arguments for better facility, better jobs, more improved communication, and so on, but all urbanization and development take place at the cost of environment which plays an important role in human well-being. The emphasis is almost always on more fast growth, while environmental impacts and the rights of nature are left out. This skewed approach needs to be engaged with the dilemmas and trade-offs required to maximize environmental justice keeping in mind the sustainability of city.

To conclude, globalization era has led to increase in incomes, while there has been urban environmental stress having a disadvantageous edge so far the well-being is concerned. SWB is the ultimate aim of every individual, hence it is very important to focus on the surrounding natural environment which is an important indicator of quality of life. This study can provide a portrait of life in an urban scenario where beside the financial matters, role of other behavioural and environmental factors have also been studied. Ultimately, money is not everything, but the subjective happiness with the surroundings environment and society also count a lot. The interactions between the two help in constituting a superior quality of life. Finally, the partnership to include environmental and behavioural professionals with local authority is important and a key to restore and improve the city environment of Bhopal for positive environment appraisal and high subjective well-being. A bright future culminating in nature led development is anticipated.

REFERENCES

Adams, J.S. (1963) "Toward an Understanding of Inequity", *Journal of Abnormal and Social Psychology*, 67, 422–36.

Aneshensel, C.S. and Sucoff, C.A. (1996) "The Neighbourhood Context of Adolescent Mental Health", *Journal of Health and Social Behaviour*, 37, 293–310.

Averill, J.R. (1973) "Personal Control over Aversive Stimuli and It's Relationship to Stress", *Psychology Bulletin*, 80, 286–303.

Barker, R.G. (1968) "Concepts and Methods for Studying the Environment of Human Behaviour", *Ecological Psychology*, Stanford, CA, Stanford University Press.

Baum, A., and Paulus. P. (1987) "Crowding", In D. Stockols and I. Altman (eds.), *Handbook of Environmental Psychology*, New York, Wiley.

Baum, A. and Singer, J.E. (eds.) (1982) *Advances in Environmental Psychology*, Vol. 4, Hillsdale, N.J., Erlbaum.

Baumeister, R.F. and Leary, M.R. (1995) "The Need to Belong: Desire for Interpersonal Attachments as a Fundamental Human Motivation", *Psychological Bulletin*, 117, 497–529.

Bell, C.C. (1987) "Preventive Strategies for Dealing with Violence Among Blacks", *Community Mental Health Journal*, 23(3), 217–28.

Bell, P., and Greene, T. (1982) "Thermal Stress: Psychological Comfort, Performance and Social Effects of Hot and Cold Environments". In G.W. Evans, (ed.), *Environmental Stress*, 75–104, New York, McGraw-Hill.

Boyden (ed.) (1970) *The Impact of Civilisation on the Biology of Man*, 219–30, Toronto, Canada, University of Toronto Press.

Cambell, J. (1983) "Ambient Stressors", *Environment and Behaviour*, 15, 55–380.

Campbell, C. and Schwarz, D.F. (1996) "Prevalence and Impact of Exposure to Interpersonal Violence among Suburban and Urban Middle School Students", *Pediatrics*, 98(3), 396–402.

Charnetski, C. and Brennan, F. (2001) *Feeling Good is Good for You*. Rodale Press.

Cochrane, R. (1983) *The Social Creation of Mental Illness*, London, Longman.

Cohen, S. and Weinstein, N. (1982) "Non-auditory Effects of Noise on Behaviour and Health", *Journal of Social Issues*, 37, 36–70.

Cohen, S., Glass, D. and Philips, S. (1979) "Environment and Health". In H.E. Freeman, S. Levin, and L.G. Reeder (eds.) *Handbook of Medical Sociology*, 134–49, Englewood Cliffs, N.J.

Crowell, B.A., George, L.K., Blazer, D. and Landerman, R. (1986) "Psychosocial Risk Factors and Urban-Rural Difference in the Prevalence of Major Depression", *British Journal of Psychiatry*, 149, 307.

de Sherbinin, A. (2003) "The Role of Sustainability Indicators as a Tool for Assessing Territorial Environmental Competitiveness", *International Forum for Rural Development*, Brasilia, Brazil, November 4–6.

Dean, K. and James, H. (1984) "Depression and Schizophrenia in an English City" in H. Freeman (ed.), *Mental Health and Environment.* London, Churchill Livingstone.

Deardorff, J., Gonzales, N.A. and Sandler, I.N. (2003) "Control Beliefs as a Mediator of the Relation between Stress and Depressive Symptoms among Inner-City Adolescents", *Journal of Abnormal Child Psychology*, Vol. 31.

Diener, E. and Lucas, R. (1999) "Personality and Subjective Well-being". In Kahneman, D., Diener, E. and Schwarz, N. (eds.), *Well-being: The Foundations of Hedonic Psychology*, pp. 213–29, New York, Russell Sage Foundation.

DosReis, S., Zito J.M., Safer D.J. et al., (2001) "Mental Health Services for Youths in Foster Care and Disabled Youths", *American Journal of Public Health*, 91, 1094–99.

Epstein, Y. (1982) "Crowding Stress and Human Behaviour". In G.W. Evans (ed.), *Environmental Stress,* 133–148, New York, Cambridge University press.

Esser, A.H. (1974) *Environment and Mental Health, Science, Medicine and Man,* 1, 181–193.

Evans, G.W. and Jacob, S,V. (1982) "Air Pollution and Human Behaviour". In G.W. Evans (ed.), *Environmental Stress*, pp. 105–132, New York, Cambridge University Press.

Folkman, S. (1997) "Positive Psychological States and Coping with Severe Stress", *Social Science and Medicine*, 45, 1207–1221.

Folkman, S. and Moskowitz, J.T. (2000) "Positive Affect and the Other Side of Coping", *American Psychologist*, 55(6) 647–654.

Garbarino, J. and Kostelny, K. (1992) "Child Maltreatment as a Community Problem", *Child Abuse* Negl.; 16:455–64.

Glass, D. and Singer, J. (1972) *Urban Stress*,. New York, Academic.

Guerra, N.G., Huesmann, L.R., Tolan, RH., Van Acker, R. and Eron, L.D. (1995) "Stressful Events and Individual Beliefs as Correlates of Economic Disadvantage and Aggression Among Urban Children", *Journal of Consulting and Clinical Psychology*, 63, 518–528.

Gump B., Matthews K.A. and Raikkonen, K. (1999) "Modeling Relationships Among Socioeconomic Status, Hostility, Cardiovascular Reactivity, and Left Ventricular Mass in African American and White Children", *Health Psychology*, 18, 140–150.

Gupta, M. and Sharma, K. (1996, third quarter), "Environmental operations Management: An Opportunity for Improvement", *Production and Inventory Management Journal*, 40–46.

Halpren, D. (1995) *Mental Health and the Built Environment*, London, Taylor and Francis.

Heilbrun, A.B. and Friedberg, E.B. (1988) "Type A Personality, Self-Control, and Vulnerability to Stress" *Journal of Personality Assessment*, Vol. 52, No. 3, 420–33.

Kaplan, S. and Kaplan, R. (1982) *Cognition and Environment: Functioning in an Uncertain World*, New York, Praeger.

Khuntia, Gayatri and Rishi, Parul (2007) An Assessment of Urban Environmental Stress and Subjective well-being of Bhopal City Dwellers, Unpublished M.Phil Dissertation, IIFM, Bhopal.

Korte, C. and Kerr, N. (1975) "Response to Altruistic Opportunities in Urban and Non-urban Settings", *Journal of Social Psychology*, 95, 183–84.

Kretźmann P. and McKnight, J. (1993) *Building Communities From the Inside Out: A Path Toward Finding and Mobilizing a Community's Assets*. Chicago, IL, Kretzmann and McKnight.

Lantane, B. and Darley, J.M. (1968) "Group Inhibition of Bystander Intervention in Emergencies", *Journal of Personality and Social Psychology*, 10, 25–221.

Lazarus, R.S. and Cohen, J. (1977) "Environmental Stress". In J. Wohlwill and I. Altman (eds.), *Human Behaviour and Environment*, 90–127, New York, Plenum.

Lazarus, R.S. (1999) *Stress and Emotion: A New Synthesis*, New York, Springer.

Lazarus, R.S. and Folkman, S. (1984) *Stress Appraisal and Coping*. New York, Springer.

Levy, L. and Rowitz, L. (1973) *The Ecology of Mental Disorder*, New York City, Human Sciences Press.

Lewin, K. (1951) *Field Theory in Social Science*, New York, Harper.

Manning, V., Best, D., Rawaf, S., Rowley, J., Floyd, K. and Strang, J. (2001) "Drug use in Adolescence: The Relationship between Opportunity, Initial Use and Continuation of Use of Four Illicit Drugs in a Cohort of 14–16 Year-olds in South London", *Drugs: Education, Prevention and Policy*, 8, 397–405.

McLeod, J.D. and Edwards, K (1995) "Contextual Determinants of Children's Responses to Poverty", *Social Forces*, 73:1487–1516.

Nespor, K. (1987) *Various Relaxation Techniques and their Comparison with Autogenic Training*, Dom techniky Bratislava (Czechoslovakia) 10:24–40, (in Czech).

Ostfield, M. and D'Atri, D.A. (1977) "Psycho Physiological Responses to the Urban Environment", *Psychosomatic Medicine*, New York, Oxford University Press, 308–21.

Paulus, P.B. (1980) "Crowding". In *Psychology of Group Influence*. Hillsdale, N.J., Erlbaum.

Ruback R. and Pandey J. (2002) "Mental Distress and Physical Symptoms in the Slums of New Delhi: the Role of Individual, Household, and Neighbourhood Factors", *Journal of Applied Social Psychology*, 32.

Safyer, A.W. (1994) "The Impact of Inner-City Life on Adolescent Development Implications for Social Work", *Smith College Studies in Social Work*, 64, 153–67.

Seligman, M.E.P. (1975) *Helplessness*, San Francisco, W.H. Freeman.

Sharma, S., Sood, N. and Spielberger. C.D. (1998) "Occupational Stress, Anxiety, Anger and Type-A Behaviour". In D.M. Pastonjee, U. Pareek and R. Agrawal (eds.), *Studies in Stress and its Management*. Delhi, Oxford and IBH.

Shumow, L, Vandell, D.L. and Posner, J. (1998) "Perceptions of Danger: a Psychological Mediator of Neighbourhood Demographic Characteristics", *American Journal of Orthopsychiatry* 68, 468–78.

Sundstrom, Eric (1978) "Crowding as A Sequential Process: Review of Research on the Effects of Population Density on Humans". In *Human Response to Crowding*, Andres Baum and Yakov Epstein, (eds.), Hills-dale, NJ, Lawrence Erlbaum.

Swami Satyananda (1980) *Sure Ways to Self Realisation*, Bihar School of Yoga, Monghyr, p. 71.

Veitch, R. and Arkkelin, D. (1995) *Environmental Psychology: An Interdisciplinary Perspective.*

Walker, B. Jr., Goodwin, N.J. and Warren, R.C. (1992) "Violence A Challenge to the Public Health Community", *Journal of National Medical Association* 84(6), 490–96.

Wong, P.T.P. (1993) "Effective Management of Life Stress: The Resource-Congruence Model", Conference Proceedings, Stress Medicine, 9, 5–60.

Part II

Environmental Attitudes and Behaviour

EDITOR'S INTRODUCTION

Human activities can have substantial impacts on the physical environment that provides for life. With the growing human population, resource management and environmental conservation have become the critical problems. Many have suggested that the roots of environmental problems stem, in large part, from the basic values upon which society has been built (Swan, 1971), such as individualism, materialism, institutions of government, progress and growth. (Christensen and Norgard, 1976; Harblin, 1977). These values may have seemed appropriate in a period of low population level but could prove disastrous for society as the human population reaches the ecological limits of the earth (Gigliotti, 1990).

Meadows, Meadows and Randers (1992) state that human use of essential resources has already surpassed sustainable rates. With the continuing rise in population, there just will not be enough resources to meet the demands. "The capacity of global life support system to sustain a technologically advanced civilization is likely to collapse within the foreseeable future" (Malone and Corell, 1989; Jacobson and Price, 1991). As Browne (1991: 91) wrote "I cannot imagine even a nuclear war as destructive to wildlife as the effects of unchecked human reproduction". He stated that humanity is mindlessly waging war against other living things in the name of "living space" and "it is a war in which victory will impoverish all of us beyond belief".

It has been well accepted by now that 'people are the key' to manage the environment in a sustainable way. In view of the mindset that people have towards environment and how the mind-set or attitudes govern and guide their behaviour is very crucial. This section on environmental attitudes and behaviour explores various dimensions of people-environment relationship and aspects of co-existence between people and environment with special reference to people-forest interface. It is true that people are highly concerned about environment but their concrete actions save the environment are yet to be strengthened. A need for strong motivation and sense of responsibility marked with internal locus of control is stressed to generate the positive mindset towards environment which has the potential to generate positive action too.

5

Mind Your Mindset

Environment Demands Positive Action

PARUL RISHI

Since 1950, while the world's population has doubled, the energy use has quadrupled, vastly improving material conditions of life for many (Jacobson and Price, 1991). However, economic growth has not been uniform throughout the world. The high level of consumption by rich countries and their great disparities with poor countries cannot continue much longer. With the continuing rise in both the population and the desires of the *have nots,* there just will not be enough resources to meet the demands.

The human activities that interact with the earth's natural systems are driven by three fundamental factors: the number of human beings and their distribution around the globe; human needs and desires, which provide individuals and societies with motivations to act; and the cultural, social, economic, and political structures that shape and mediate their behaviour (Jacobson and Price, 1991).

MINDSET FOR ENVIRONMENT: THE ATTITUDES

In view of the above background, there is nothing as much required as the positive mindset or attitudes to nurture the mother nature with concern and care for its sustainable future. The study of environmental attitudes, together with attitudes concerning the modalities for evaluating 'environmental qualities', constitute the

Figure 5.1 Environment Behaviour Relationship and Positive Environmental Appraisal

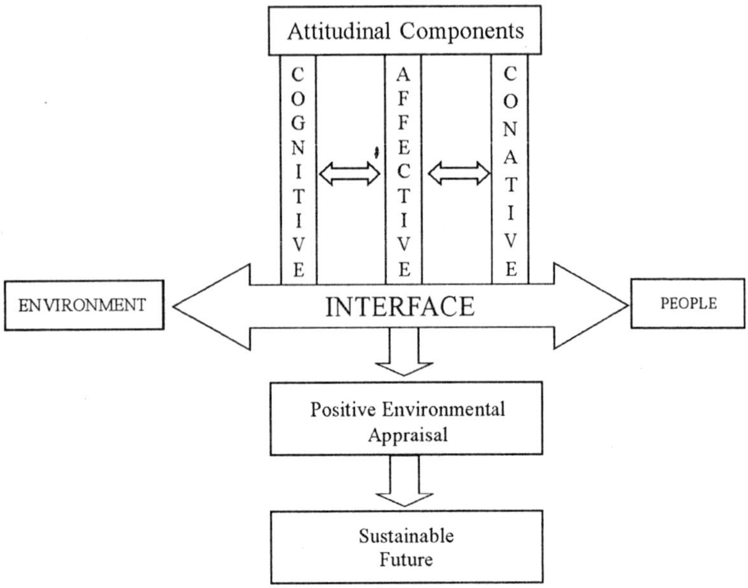

area of research most directly involved with the issue of satisfaction/dissatisfaction with the characteristics of various types of environments. With regard to attitudes, specific attention has been given to the importance of the positions that individuals take concerning the most current and relevant problems and risks, such as pollution and environmental deterioration. 'Attitude' is a concept used by social psychologists to account for variations in individual behaviour in similar circumstances. When concerned with 'environmental attitudes', the research focuses attention on the affective-evaluative and cognitive-informative content of the positions persons adopt towards aspects or specific properties of the physical environment (Figure 5.1). As Stokols (1978) pointed out, in this case the main objective consists in showing tendencies to respond favourable/unfavourably, to the environmental characteristics considered.

Another aspect also characterizing this study is its quantitative approach and the extensive use of measurement instruments, such as 'attitude scales'. Empirically speaking, this approach follows two main directions: the first considers attitudes as expressions of individual satisfaction/dissatisfaction towards the environment, and it primarily emphasizes the evaluative component; the second tends to approach them with respect to several important environmental problems (such as pollution, limited resources and environmental risk), and pays particular attention to the informative component (Stokols, 1978).

For the effective conservation and management of environment, it is very relevant that human beings should have positive beliefs, feelings and attitudes about environment. This positive environmental appraisal provides guidelines for predicting their behaviour towards environment in different situations. In this regard, attitudinal assessment becomes very important part of environmental conservation and management.

Attitudinal Framework

The understanding of attitudes is one of the central concerns in social life. It is a relatively permanent enduring state of readiness of mind, which predisposes an individual to react in a characteristic way to any event, situation or person with which it is related. Krech, Crutchfield and Ballachey (1962) indicate that social actions of the individual reflect his attitudes, which are the enduring systems of positive or negative evaluations, emotional feelings and action tendencies with respect to social objects. The tri-componential viewpoint represents a paradigm of people-environment relationship in their attitudes in the following way:

The way people are treating natural environment is determined by their beliefs and feelings about it, and their dispositions to respond to them. All the environmental actions of people are directed by their attitudes. By assessing their attitudes, it is possible to do something about prediction and control of their behaviour which may be ultimately useful for resolving people-environment conflicts and ensuring peaceful co-existence for conservation, management and development. If we know the attitudes of people towards environment, its resources and conservation, they will be better able to find out the strategies to promote pro-environment behaviour in a more meaningful way.

Attitudinal Measurement

The task of measuring attitudes is not a simple one. Henerson et al. (1987) state that the concept of attitude, like many other abstract concepts, is a creation—a construct. It is a tool that serves the human need to see order and constancy in what people say, think and do, so that given certain behaviours, predictions can be made about future behaviours. An attitude can not be measured the way heart beat can be measured... we can only infer that a person has attitudes by his/her words, actions or statements" (Henerson et al., 1987).

Behaviour is a result of many complex factors—feelings, past experiences, assessment of expectations of others, anticipation of consequences and so on. The patterns of behaviour are also not very consistent. But still, a person's statements about his own attitude, are the best indicators of the attitude, particularly when there is no reason to hide anything.

Keeping in view the precautions given in Box 1, various rating scales, questionnaires and interview schedules can be designed to measure the attitudes and related constructs. Besides, narrative technique is also important in the understanding of environmental attitudes (Shanahan and McComas, 1999

Box 1

- Precautions for attitudes measurement
- When we measure attitudes we must rely on inference, since it is impossible to measure attitudes directly as it is a subjective construct.
- Behaviours, beliefs and feelings will not always match, even when we correctly assume that they reflect a single attitude; so to focus on only one manifestation of an attitude may tend to distort the picture of situation and misleads.
- We have no guarantee that the attitude, we want to assess will be reliable long enough for a one time measurement. A fluctuating attitude cannot be revealed by the information gathered on one occasion.

Source: Henerson et al., 1987.

Shanahan, Pelstring and McComas, 1999). The significance of researchers' observations too, however, can not be undermined.

The Value-Attitude-Behaviour Model

Debate regarding how environment should be managed often occurs because different interest groups hold differing values. Environmental activists are often at odds with those who value the consumerist use of environment and natural resources. To a large extent, the debates centre on preservation vs use-value orientations (Stern and Dietz, 1994). Conflicts in values have increased because utilitarian views are being replaced by environmental orientation (Stern et al., 1995). As these trends are likely to continue in the future, it is important for people to understand how values influence attitudes and support/oppose environmental management policies. Most of the research has presented correlational evidences to support the relation between values, attitudes and behaviour.

The value-attitude-behaviour framework suggests that an individual's view of the environment can be organized into a

Figure 5.2 The Hierarchy Model of Human Behaviour (adapted from Fulton et al., 1996)

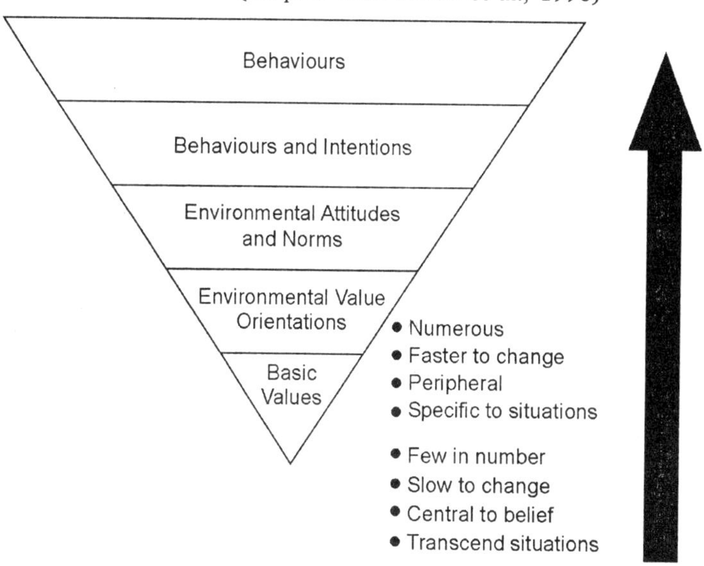

cognitive hierarchy consisting of values, value orientation, attitudes/norm, behavioural intentions and behaviours (Fulton et al., 1996). Each of these elements builds upon one another to form an inverted pyramid (Figure 5.3).

There is, however, the need to expand the cognitive hierarchy model to include the determinants of values and value orientations. Previous studies have suggested (Manfredo and Zinn, 1996) that values differ by respondents' age, gender, location of residence (urban/rural), political orientation and membership in environmental organizations. Further research is required to examine such determinants and clarify their relationships. Vaske and Donnelly (1999) have demonstrated that the value, attitude-behaviour model provides a useful theoretical framework for addressing environment and natural resource management issues. The importance of attitudinal studies has been increased owing to incorporation of the human factor in conservation and management of environmental resources.

ASPECTS FOR CO-EXISTENCE: ENVIRONMENT AND PEOPLE

Extensive research has been conducted to assess the environmental attitudes of people in different natural surroundings. Nepal and Weber (1995) present the findings of a field survey on the interface between local people and wildlife in the Royal Chitwan National Park, to explore the potential co-existence of wildlife and local people. Based on three distinct areas defined by their distance to the park, the complexities and dynamics of local people's perceptions, attitudes and motivation towards wildlife conservation were analysed.

The findings indicate that although local people disliked the restrictions imposed by the park authority, curtailing the use of park resources, they take a positive attitude to wildlife conservation. Their active involvement in protection and conservation could be secured if due consideration is given to their needs which have to be addressed in the overall planning and management of the park.

An assessment of environmental knowledge and attitudes held by community leaders of Taiwan was conducted by Shih-Jang Hsu and Robert Roth (1996). Significant differences were found in knowledge scores by age, education level, income level and ethnicity and in attitude scores by age, educational level, occupation and ethnicity. The results indicate that education level is the best predictor of environmental knowledge and attitudes. By increasing the knowledge about environmental awareness and attitudes. By increasing the knowledge about environment through education, we can expect a more positive attitude among people towards environment and natural resources.

Pimbert and Gujja (1997) in their paper, "Village Voices Challenging Wetland Management Policies: Experiences from India and Pakistan" indicated that effective management of two wetland areas of international importance for conservation of Keoladeo National Park in India and Ucchali wetland complex in Pakistan—has been hampered by local opposition to existing wetland management policies and the threat of worsening conflict between villagers and conservation authorities. Participatory Rural Appraisals (PRA) reveal that strict restrictions on the use of park resources by local communities, such as the banning of grazing in Keoladeo, were ecologically unsound as well as a threat to local livelihood. The greatest challenge for policy makers is for PRAs to be conducted with local communities before protected areas are formally established. Such measures would encourage joint management, reduce potential conflicts and promote ecologically sound practice.

Another empirical study on values, ethics, and attitudes towards national forest management by Manning Valliere and Minteer (1998), who measured environmental values and ethics and explored their relationships to attitudes toward national forest management. The principal research methods were literature review and a survey of Vermont residents concerning management of Green Mountain National Forest. Descriptive findings suggest that respondents favour nonmaterial values of national forests, subscribe to a diversity of environmental ethics, including anthropometric and bio/ecocentric, and support emerging concepts of ecosystem management. Environmental values and ethics explain approximately sixty per cent of the variation in attitudes towards national forest management.

In rethinking the decentralization and devolution of biodiversity conservation as a step towards positive environmental attitudes, Enters and Anderson (1999) have stated that it is essential to look at community-based management with a critical eye. A community-centred approach to biodiversity conservation may be just as unsatisfactory as completely government-control approaches. The best solutions for the future are likely to be found in partnerships and dynamic interactions among different stakeholders.

Parul Rishi (1999) has conducted a behavioural analysis of local people and other stakeholders in Keoladeo National Park (KNP). The conservation attitudes and perception of local villagers, tourists and local city dwellers were assessed using the conservation attitude scale. The behavioural analysis of situation revealed that there was a persisting conflicting situation in KNP as far as the people-KNP interface is concerned, with special reference to grazing and fodder. However, ecological importance of the national park is well recognized and unquestioned by all the stakeholders. Local villagers as well as city dwellers of Bharatpur have strong feeling of affinity with KNP, that is why, its management cannot remain untouched by their influence.

People in and around Gir National Park exert tremendous biotic pressure for grass, firewood and small timber in particular. Since these are prime necessities of people and livestock and in the absence of acceptable alternative, people, especially the poor and compelled to adopt illegal ways for obtaining them. Besides, there are other issues also like, problem of trust, difficulties of communication, stakeholder' number, power differences, degree of risk and uncertainty etc., which need attention.

The environmental issues are governed by a web of interests, conflicts and trade-offs between different sets of local people, government departments and local planners (Shrivastava, 1997), which are mentioned as under:

- Divergent interests competing over the same resource.
- Religious favour overpowering conservation.
- Restriction on use of roads passing through Gir National Park.
- The problem on Gauchar (common grazing areas).
- NTFP collection.
- Agriculture vs forestry interest.

- Damage by wildlife.
- Tourists' expectations.
- Local press and wildlife.

Looking at the above issues and conflicts, the attitudinal assessment of people about the park and its resources became a very pertinent need.

ENVIRONMENTAL EVALUATIONS

The evaluations that people express about the various properties of the environmental settings in respect of their daily life takes place, especially with reference to residential and/or urban settings, constitute the premise first towards grasping the importance or various properties or components in defining the degree of environmental satisfaction/dissatisfaction, and, secondly, for verifying the influence that can be exercised on these evaluative hierarchies by socio-demographic variables (such as sex or age) and/or by some other variables that contribute towards specifying the more complex relationship between persons and the environment (for example, socio-economic level, working or not, length of environmental residence).

Stokols (1981) suggests, environmental evaluation is a function of the 'congruence' people perceive between their own needs and aim and the opportunities offered by the environment, rather than as a response linearly attributed to the objective properties of the latter. The attitudes expressed by residents about their neighbourhood consist in two main tendencies regarding:

1. The central position assumed by some particular environmental characteristics in determining the degree of positive connotation of the attitudes;
2. The connections linking the evaluations with length of residence and time spent in the neighbourhood everyday.

ENVIRONMENTAL PROBLEMS DEMAND HUMAN CONCERN

The increasing interest of environmental psychologists in the study of attitudes towards some environmental problems is linked to the growing objective relevance that these problems have assumed.

Specific problems include the intensive exploitation of resources, the various forms of environmental pollution, and risks that some systems of energy production (primarily nuclear) can involve for people's health and well-being. Empirically speaking, research has been primarily oriented towards the study of ways in which people perceive and evaluate these problems. Particular attention has been devoted to the types of information individuals have about specific problems, since information is their means of reference for structuring attitudes. Stokols (1978) has also proposed verifying the coherence between attitudes and behaviours people tend to adopt in view of an improvement in environmental conditions.

The study of environmental attitudes also includes the formulation of predictions about behaviours. The literature shows the prevalent tendency to primarily focus on the degree and quality of information/knowledge people have about various environmental problems. That is the starting point from which investigations are made to point out the connections between information, people's evaluative positions towards specific environmental problems and readiness of behaviours for safeguarding and improving environmental conditions.

Resource managers and conservationists are generally trained in natural sciences so they concentrate their efforts on biological and technological solutions to environmental crises like reducing emissions from factories, devising fuel efficient vehicles, installing water treatment plants etc. All these are definitely the valuable contributions toward conservation and management, but technology alone cannot solve the environmental problems. Ecological crises is much more a problem of maladaptive behaviour than just a technical problem. The cumulative effect of human activity is threatening the ecological systems upon which we all depend. We know that introduction of new chemicals in the products of our daily usage, failure of recycling are serious shortcomings in much of our industrialized world. Yet knowing all that, we find ourselves sucked into consumption, choosing an item more for its attractive packaging or its image than for real need it satisfies (Foxall, 1992). All aspects of the production and sale of goods—from extraction of resources to their transport, conversion into useful products and eventual disposal have serious environmental repercussions. But the real fact is that all such industries producing environmentally sensitive products and eventual disposal have serious environmental repercussions. All such industries producing environmentally sensitive products are on the go because we, humans, either do not know about their negative impact on environment or can not resist our temptation of purchasing convenient, economical and attractive products and blind our eyes from their impact on environment. That is why solutions to ecological crises require a behavioural approach to environmental conservation and management.

The human activities that interact with the earth's natural systems are driven by three fundamental factors: the number of human beings and their distribution among the globe, human needs and desires, which provide individuals and societies with motivations to act; and the cultural, social, economic and political structures that shape and mediate their behaviour (Jacobson and Price 1991).

MIND YOUR MINDSET FOR PRO-ENVIRONMENTAL BEHAVIOUR

A study on college students of Bhopal city by Parul Rishi (2001) has revealed that they were highly concerned about and well aware of environmental problems, i.e., they have a strong pro-environmental attitude, but on the other hand, their actual commitments towards pro-environmental activities were quite limited. They readily made verbal commitments for reducing air pollution, conserving environmental resources or reducing waste but their actual behaviour was inconsistent with commitments. These results can be explained, first, with reference to attribution style, and secondly with reference to attitude behaviour inconsistency studies. It is a common human tendency to safely assume that it is always the situation or the other person, who is the culprit. It is known as external locus of control. People tend to blame industries or government for local, national and global environmental problems. As the present study was conducted in Bhopal city, the environmental might have been also expectedly influenced by the great environmental disaster known as "Bhopal

Gas Tragedy" which occurred at Union Carbide in 1984. The people interviewed were between 6 and 14 years of age at the time of gas tragedy which is a very critical age for any trauma and its significant effect on the mind. Eighty-eight per cent of them was residing in Bhopal itself at the time of the tragedy. The group's attitudes could have been significantly influenced by the traumatic experiences at that time and the post-traumatic stressful experiences, physiological and psychological.

In a study by Crowley and Hoyer (1994), 75 per cent of consumers said that business is at fault for not developing consumer products that are eco-friendly. In another study, 70 per cent consumers admitted that they are more interested in conveniences than in environment (Waldrop, 1991). On the other hand, a typical survey finding from 1991 *Wall Street Journal* found that more than 80 per cent of consumers find it more important to protect the environment than keeping prices down. Similar results were also found in the study by Parul Rishi (1999) where 72 per cent respondents agreed to donate a days pay to the foundation to help improve environment and 79 per cent were willing to even pay the pollution tax, if it is going to reduce the pollution problems.

FROM POSITIVE MINDSET TO POSITIVE ACTION

Before we do anything for environmental problems, there is a need to take the individual responsibility for existing environmental situations. The logic behind this is that small steps taken by us can make a big difference in situation. It is just a matter of attribution style. If people take individual responsibility of environmental crises, they feel themselves capable of controlling further harm, and it is an internal locus of control which is very helpful in promoting pro-environment behaviour. People with external locus of control offer several justifications for their behaviour, e.g., (i) if no one is behaving in a pro-environmental way, what difference my behaviour will make to the environment? or (ii) government should take strong steps. It is governments' responsibility to control environmentally damaging activities. The answer to all these excuses is 'just don't contribute' to make the problems even worse. People's concern about the environment is highly related to their attitudes and motivations. A best known segmentation of environmental attitudes was developed in 1980 by Roper Organization and S.C. Johnson and Sons. They identified five categories of people:

1. *True blue greens:* Activity concerned about environment, consistency between actual behaviour and environmental attitudes and concern, leaders of environmental movements, strong internal locus of control. This segmentation corresponds to Maslow's (1954) self-actualization, the highest hierarchy among needs.
2. *Green back greens:* Committed to environment in an idealistic sense of the term, e.g., ready to pay higher prices for green products, pollution tax, day's pay for environmental conservation etc. It corresponds to self-esteem need among Maslow's (1954) hierarchy. The high score on verbal commitment towards environment indicates this segmentation.
3. *Sprouts:* Mild concern about environment and mild behavioural response. Their pattern of pro-environment behaviour is yet to be exhibited. It corresponds to Maslow's (1954) belongingness and love needs.
4. *Grousers:* Consistently rationalize lack of pro-environmental behaviours by offering all kinds of excuses and criticizing the poor performance of others. Their locus of control is external. It corresponds to Maslow's (1994) safety needs.
5. *Basic browns:* They also have external locus of control and they do not believe that people can make a difference to environment. They do no pro-environmental effort on their part, and it is a least enviro-active group corresponding to the lowest among need hierarchy, i.e., physiological needs.

The said framework indicates that most of the people try to express positive environmental attitudes, but there is attitude behaviour inconsistency as well. People often assume that knowledge influences attitude which in turn affect behaviour, but is not always true. In Parul Rishi's study (1999), the correlation between affective-cognitive and cognitive-conative domains were moderate to low positive. Hines et al. (1987) found a moderate positive correlation between attitude and behaviour in the meta-analysis of 51 studies. However, a recognized discrepancy between attitude and behaviour has been observed in many psychological studies (Lucas, 1980; Rajecki, 1982; Hangtredt and Wegener, 1994).

ATTITUDE-BEHAVIOUR INCONSISTENCY

There are various potential reasons for attitude-behaviour inconsistency. One is the temporal instability, i.e. the interval between expression of attitude and behaviour/action. The greater the interval, the greater the inconsistency. If a persons is asked about his/her attitude towards things leading to ozone depletion, s/he replies in an eco-friendly manner but when she goes to purchase products like refrigerator or other products which have contents under pressure, she ignores the pro-environment attitude and in spite of having knowledge of CFC being used in it he decides to purchase, because considerations of price, performance and convenience become more important. To eliminate these reasons of inconsistency, specific pro-environmental attitudes should be promoted instead of the general ones. By publicizing such appeals like 'do not use non-recyclable products', or 'use public transport to conserve oil' or 'make car pools to save oil', or read the statement 'ozone friendly, contains no CFC' before purchasing the products which contain contents under pressure etc.

Normative influences also exert a strong force over behaviour. If family/society have a social norm against—pro-environment behaviour we express the same regardless of our true feelings leading to apparent attitude behaviour inconsistency (Dabholkar, 1994). So, if strong social norms for responsible environment behaviour existed, more people might behave responsibly irrespective of their less strong pro-environmental attitudes. Promotion of pro-environmental attitudes should start from informational educational campaigns followed by explicit demonstration of undesirable behaviour and its consequences on the media (modelling) followed by reinforcement techniques like providing incentives for pro-environmental behaviour. If all the three are not much effective then fear appeals should be publicized to arouse the people emotionally and then recommending pro-environmental behaviour, e.g., 'save energy before we run out'; 'degradation of forest will end the life'; 'smoke and go to cancer hospital'; 'vehicular pollution decreasing life span, etc.

CONCLUSION

To conclude, people are effectively concerned and intellectually aware about environmental problems and are ready to commit responsible behaviour towards environment but their actual commitment of pro-environmental activities is woefully low. Most of the people fall within the category of green back greens who are idealistically concerned with environment to satisfy their need for self-esteem, i.e, they are able to mind their mindset but are not that strongly geared up to act in a pro-environmental way as they should be. Therefore, promoting pro-environmental behaviour is the need of the present era if we want to save the earth. Ultimately human action will be the basis for success or failure of sustained diversity of life. The factors identified as having positive influence on pro-environmental behaviour are internal locus of control, sense of responsibility, and positive attitude. Promoting all these factors, first of all, requires strong motivation on the part of initiator as well as the target persons or group. If we are able to create a society in which people can understand the implications of all their actions for themselves, the rest of the living environment and for future generations, our objectives are accomplished, because ultimately, the need to be able to make their own moral decisions about environmental, matters the most. So, one has to mind the positive mindset for environment, but it should be clubbed with the positive and concrete pro-environmental action too.

REFERENCES

Browne, Malcolm W. (1991) "Living on Borrowed Time", *The New York Times*, December 15.

Christensen, B. and Norgard, J. (1976) "Social Values and the Limits to Growth", Cited in 'The Limits and Beyond: Global Collapse or a Sustainable Future?' by Doug J. Porter, *Sustainable Development*, 1(2), pp. 53–67.

Crowley A.E. and Hoyer W.D. (1994) "An Integrative Framework for Understanding Two-sided Persuasion", *Journal of Consumer Research*, 21, pp. 561–574.

Dabholkar, Prathiba (1994) "Incorporating Choice into an Attitudinal Framework: Analysing Models of Mental Comparison Processes", *Journal of Consumer Research*, 21 (1), pp. 100–118.

Enters, T. and Anderson, J. (1999) "Rethinking the Decentralization and Devolution of Biodiversity Conservation", *Unasylva*, 50 (4), pp. 6–11.

Foxal, G.R. (1992) "The Behaviroural Perspective Model of Purchase and Consumption: From Consumer Theory to Marketing Practice", *Journal of the Academy of Marketing Science*, 20(2), pp. 189–198.

Fulton, D.C., M.J. Manfredo, and J. Lipscomb. (1996). "Wildlife Value Orientations: A Conceptual and Measurement Approach", *Human Dimensions of Wildlife* 1, pp. 24–27.

Gigliotti, L.M. (1990) "Environmental Education: What Went Wrong?" *Journal of Environmental Education*, 22, pp. 9–12.

Harblin, T.D. (1977) "Mine or Garden? Values and the Environment: Probable Sources of Change in the Next Hundred Years", *Zygon*, 12(2), pp. 134–150.

Henerson, M.E., Morris, L.L., Jayalor C. and Gibbon F. (1987) *How to Measure Attitudes*? California, Sage Publication, Inc.

Jacobson, Harold K. and Price, Martin F. (1991). *A Framework For Research On The Human Dimensions of Global Environmental Change.* Barcelona, ISSC/UNESCO, p. 61.

Krech, D., Crutchfield, R.S., Ballachey, E.L. (1962) "Individual In Society", *A Textbook of Social Psychology*, New York, Mcgraw-Hill.

Malone, T.F., Corell R. (1989). "Mission to Planet Earth Revisited". *Environment,* 31(3), pp. 6–11.

Manfredo, M.J., and H.C. Zinn (1996) "Population Change and its Implications for Wildlife Management in the New West: A Case Study of Colorado", *Human Dimensions of Wildlife 1,* pp. 62–74.

Meadows, D.H., Meadows, D.L. and Randers, J. (1992) *Beyond The Limits To Growth. Dancing Towards the Future* (IC#32) Summer, Page 10, URL: http://www.context.org/ICLIB/IC32/Meadows.htm

Nepal, S.K. and Weber, K.E. (1995) "Aspects for Coexistence: Wildlife and Local People", *Ambio,* 24 (4), pp. 238–245.

Pimbert, M.P. and Jujja, B. (1997). "Village Voices Challenging Wetland Management Policies". *Nature & Resources,* 33, (1), pp. 34–42.

Rishi, Parul (1999) "A Behavioural Analysis of Keoladeo National Park", *IIFM Newsletter*, Vol. II, No. 4, pp. 4–5.

Rishi, Parul (1999). Economic and Behavioural Analysis of Protected Area Management: A Study of Keoladeo National Park, Unpublished Research Report, IIFM, Bhopal.

Shanahan and McComas (1999) *Nature Stories,* Cresskill, NJ: Hampton Press.

Shih-Jan Hsu and Roth R.E. (1996) "An Assessment of Environmental Knowledge and Attitudes Held by Community Leaders in the Hualien Area of Taiwan". *The Journal of Environmental Education*, 28, (1), pp. 24–31.

Shrivastava, A. (1997) "People's Participation: A Vital Component in Management of GIR Protected Area", *Indian Forester*, pp. 503–511.

Stern, P.C., Dietz, Thomas and Guagnano, G.A. (1995) "Values, Beliefs and Pro-environmental Action: Attitude Formation Toward Emergent Attitude Objects". *Journal Applied Social Psychology,* 25, pp. 1611–1636.

Stern, P.C. and T. Dietz. (1994) "The Value Basis of Environmental Concern". *Journal of Social Issues,* 50, pp. 65–84.

Stokols, D. (1978) "Environmental Psychology". In M.R. Rosenzweig and L.W. Porter, (eds.), *Annual Review of Psychology*, pp. 29, 253–295.

Swan, J.A. (1971) "Formation of Environmental Values: A Social Awareness", *Environment & Behavior,* 3, pp. 223–29.

Vaske, J.J. and Donnelly M.P. (1999) "A Value–Attitude-Behaviour Model Predicting Wildland Preservation Voting Intensions". *Society & Natural Resources*, 12, pp. 523–537.

6

People-Environment Relationship

Some Insights

NACHIKETA TRIPATHI

In the last two decades or so there have been many conceptualizations to understand relationship of people with their environment. Some of these conceptualizations were an attempt to find answers for the sustainable environment protection and welfare for humankind. Still, there is certain amount of ambiguity in terms of deciding which approach would be the best for lasting solution towards environment protection.

The present paper attempts to analyze some of the contemporary models of people-environment relationship (PER) and with the help of empirical findings, he has tried to throw some light on PER. In order to understand PER, it would be better to know some of the approaches suggested by other scholars. One such approach comes from Altman and Rogoff (1987), which broadly represents the kind of approach taken in the study of environmental psychology, but which can also be considered in the specific context of person-environment relationships. Four approaches to psychological analysis are described in this view, namely:

1. the *trait approach*, in which behaviour with respect to the environment is investigated in the form of objectively measurable units independently of other variables. The focus in this approach is the *person*;

2. the *interactional approach* which focuses on the combined effects of personal and environmental variables on behaviour. In this approach also, behaviour is studied in terms of objectively measurable units;
3. the *organismic approach*, which views the person-environment relationship as a whole, while acknowledging personal and environmental variables as parts of the whole. In this approach, global concepts such as the ecosystem become major independent variables; and
4. the *transactional approach*, which takes the position that the determinants of behaviour with respect to the environment can never be studied as separate units; instead, it assumes that the so-called 'parts' are built into the whole. Thus, every effect is a part of the cause, and every cause is a part of the effect, and the two co-exist in a dynamic relationship.

The second view is proposed by Stokols (1990), and is labelled as the *people-environment relationship* (PER) view. According to this conceptualization, the relationship of the individual to the environment may be seen in three different views, namely, the *minimalist, instrumental and spiritual views.* The *minimalist* view states that the physical environment has a negligible influence on human beings. In fact, often, giving importance to the environment may be detrimental to efficiency and to human growth. The *instrumental* view suggests that the physical environment is to be used as a means for achieving behavioural and economic goals. Any environment is assessed in terms of its capacity to provide resources, comfort, safety, and overall economic well-being. In other words, this view assumes that the environment exists mainly for the service of people. The *spiritual* view states that the environment exists, not for use by people, but as a medium for human values to grow and evolve. The environment is not a means to a human end, but an end in itself. Therefore, the environment must be perceived and understood, not in material terms, but in socio-cultural, spiritual or religious terms. Such aspects are reflected in the symbolic and spiritual dimensions of the environment.

The third view related to the environment is that of *environmentalist values*, proposed by Merchant (1992) and Stern and

Dietz (1994). In this approach attitudes and behaviour towards the environment are discussed in the light of certain values which are said to form the basis of environmental ethics. These are, in Merchant's view, (i) *egocentric values*, which emphasize "the maximization of self-interest by the individual", (ii) *homocentric values,* which emphasize "the greatest good for the greatest number of people", and (iii) *ecocentric values,* which emphasize the "unity, stability, diversity, and harmony" of the ecosystem.

Stern and Dietz (1994) describe three very similar values, namely, *egoistic*, *altruistic* and *biospheric*. Egoistic values correspond to Merchant's egocentric values. Altruistic values are based on the sense of obligation to others, and imply that people should work for the welfare of others rather than for oneself. Thus, one should bear in mind the consequences of one's actions for others. These values seem similar to Merchant's homocentric values. Biospheric values are similar to Merchant's ecocentric values. They assume that all components of nature, including human beings, are a part of the biosphere or ecosystem. Therefore, human actions would affect the biosphere as a whole. There is also some overlap among the views. For example, egocentric or egoistic environmentalist values seem to have something in common with Stokols' instrumental view of PER. Similarly, ecocentric or biospheric environmentalist values would possibly be associated with the spiritual view of the PER; the latter may also include homocentric or altruistic values. With respect to Altman and Rogoff's worldview approaches, it appears that the spiritual view of PER necessitates an organismic as well as transactional approach to the analysis of person-environment relationships.

Apart from above models of PER, one can find some interesting account of PER in some of the Indian writings. For example, to begin with, Indian philosophical thought sees the individual as a part of the extended environment. It acknowledges the fact that the environment (nature) is the source of all life and the provider of resources; therefore, it is not to be exploited or misused. In fact, nature is a manifestation of the divine, like the human being. Therefore, worshipping the natural environment is not merely a display of paganism, but an expression of gratitude and the acceptance of human dependence on nature. These attitudes are exhibited rather clearly in ancient Indian texts, religious practices, and in tribal and rural Indian life. Sen (1992) presents superb accounts of these aspects in her description of 'indigenous visions' of the environment. Secondly, there are some excellent examples of attempts to establish closeness with nature in human life. Shantiniketan, an educational institution in West Bengal, India, was founded by the Nobel Laureate and Indian poet, Rabindranath Tagore, with the aim of imparting education in the midst of nature. Focusing on Tagore's vision of ecological harmony, Ray (1990: 303) writes, "... the philosophy of education professed by Tagore has incorporated the idea that the role of nature is as much essential to the possibility of education proper as any other aspect". Thus the core Indian culture inculcates a PER view that suggests an essential harmony between the person and the natural environment, including other people.

At the same time, societal characteristics encourage an affiliative and community orientation (as seen in the rich Indian folk tradition). Parallel to these core cultural attitudes, economic necessity makes the Indian people instrumental in their attitude towards natural resources, as is the case in other cultures. In addition, this may give rise to a symbiotic view of PER, that is, people and nature are meant for each other. All of these possibilities taken together may give rise to a paradoxical PER view. On one hand, a spiritual view of PER goes against an instrumental or exploitative view of the environment, on the other, necessity encourages an instrumental view. Which of these PER views should predominate? The matter needs to be empirically examined.

In order to test the above theoretical ideas, an empirical study was conducted to find out what kind of PER exists in our society. For this purpose, an empirical study was designed and the details of this study (Tripathi, 2003) are given below.

METHOD

Subjects

Eighty urban respondents (40 males and 40 females) participated in the study. The sample was taken from an industrial city in northern India. The age of the male respondents was between 25 and 53 years (Mean = 39.5 years, SD = 8.61), and that of the female respondents, between 25 and 45 years (Mean = 33.15 years, SD = 7.05).

Procedure

All respondents were interviewed at their residences. The interview schedule consisted of an inventory (the PER view inventory) specifically devised for the purpose of the study. All interviews were conducted in Hindi, the native tongue of the respondents.

Measures

The PER view inventory: In view of the theoretical ideas discussed above, Stokols' (1990) conceptualization of the PER view was used as the base for the construction of the inventory. In the case of the PER view inventory, in addition to the minimalist, instrumental and spiritual views, a fourth component, namely, symbiosis, was included. The reason of including symbiosis was that it is possible that people might perceive their relationship with the environment in a reciprocal way, that is both are mutually benefited. Such dimension is not a part of Stokols' (1990) conceptualization of PER view. Statements were selected to represent each of these components (three items in each component, with a total of 12 items to measure the four PER view components). Respondents were asked to indicate the extent of their agreement or disagreement with each statement on a 7–point Likert-type scale—1 indicating strong disagreement, 7 indicating strong agreement. In addition to the PER statements, two other items were added in order to obtain information about respondents' view of the environment. The final PER view inventory thus consisted of the following items:

1. A question pertaining to the meaning of the environment (open-ended);
2. The extent to which the environment was considered important (to be rated on a 7–point scale, 1 indicating 'Not at all important', and 7 indicating 'Very important'), and reasons for the same (open-ended);
3. Twelve statements pertaining to the four PER components (described above): the respondent had to rate the agreement/disagreement with each statement (on a 7–point scale, as indicated above).

For the purpose of factor analysis, the ratings given in response to the 12 items related to the four PER components were considered.

RESULTS

The first step in the analysis of the interview responses was to compare males and females. As indicated by the means and standard deviations presented in Table 6.1, overall, there were some significant gender differences with respect to certain items. However, a closer inspection of these differences revealed that they were not of such a magnitude that one could argue for real gender differences in the two variables. Moreover, finding gender differences was not the main aim of this study. Therefore, in the interests of sample size and generalization of the findings, the remaining analysis was carried out on the combined sample of males and females.

Table 6.1 Mean and Standard Deviations of People-Environment Relationship (PER) View (N=80, 40 males, 40 females)

Item	Males		Females		Male/female comparison 't'	
	MEAN	SD	MEAN	SD		
1. PER 1 (Minimalist)	2.22	0.89	1.47	0.50	4.63	P < .01
2. PER 2 (Minimalist)	2.30	0.51	1.82	0.59	3.82	P < .01
3. PER 3 (Minimalist)	3.40	0.84	2.50	1.28	3.71	P < .01
4. PER 4 (Instrumental)	6.15	0.36	6.40	0.77	–1.84	NS
5. PER 5 (Instrumental)	3.87	0.79	2.57	1.59	4.61	P < .01
6. PER 6 (Instrumental)	5.80	1.56	6.17	1.40	–1.12	NS
7. PER 7 (Spiritual)	4.27	0.71	5.40	1.39	–4.55	P < .01
8. PER 8 (Spiritual)	4.47	1.06	5.12	1.18	–2.59	P < .05
9. PER 9 (Spiritual)	4.82	0.90	4.72	1.71	0.33	NS
10. PER 10 (Symbiotic)	5.65	0.48	5.70	1.06	–0.27	NS
11. PER 11 (Symbiotic)	3.60	0.77	4.10	1.54	–1.82	NS
12. PER 12 (Symbiotic)	4.30	1.26	4.27	1.94	0.07	NS

The meaning of environment: The responses given to this item could be classified into (a) geographical and natural aspects, such as trees and plants, rivers, mountains and so on (100%), (b) man-environment aspects (indicating that the human being is a part of the environment) (18.75%), and (c) man-made objects, such as buildings, factories and the like (18.75%). In other words, respondents perceived the environment mainly in physical terms, with a predominance of natural components.

Importance of the environment: The mean rating of importance of the environment indicated that in general, the environment was considered highly important (Mean = 6.17, SD = 0.49). Among the reasons for considering the environment important, all respondents mentioned that the environment was necessary for their existence, and for the fulfillment of needs, indicating a utilitarian or instrumental attitude.

Components of PER view: Factor analysis (using principal components method with varimax rotation) of the responses to the main PER items was carried out. This yielded four major factors which were labeled as follows: utilitarian attitude, interactiveness/symbiosis, negative orientation, and spiritualism. The factor loadings of the constituent items of the four factors are presented in Table 6.2. The first factor, namely, utilitarian attitude, included 1 item of the original minimalist component (negatively loaded), 2 items of the original instrumental component, and 1 item of the original symbiosis component. The second factor, namely, interactiveness/symbiosis, consisted of 1 item of the original minimalist component (negatively loaded), 2 items of the original spiritualism component, and 2 items of the original symbiosis component. The third factor, namely, negative orientation, included 2 items of the original minimalist component, 1 item of the original instrumental component, and 1 item of the original spiritualism component (negatively loaded). The fourth factor, namely, spiritualism, consisted of 1 item of the original minimalist view (negatively loaded), and 2 items of the original spiritualism component.

It may be noted that the PER view factors obtained in the present study were similar, but not identical, to those proposed by Stokols (1990). In order to answer the question regarding the predominant component of PER view, the means of the ratings with respect to each factor were compared. It was observed that the mean rating was highest in the case of PER Factor 1, utilitarian attitude (Mean = 5.22, SD = 0.40) and PER Factor 4, spiritualism (Mean = 5.17, SD = 0.69), followed by PER Factor 2, interactiveness/symbiosis (Mean = 4.56, SD = 0.92); it was lowest in the case of PER Factor 3, negative orientation (Mean = 2.99, SD = 0.46).

Table 6.2* People-environment Relationship (PER) View Factors and Their Loadings (N=80)

PER Factor 1: Utilitarian Attitude	
Items	**Loadings**
PER 3 (Minimalist)	–0.46
In our country the way people show their carelessness towards the environment, it appears that they do not have any concern for environment.	
PER 4 (Instrumental)	0.84
Environment is basically/primarily meant for human beings, and therefore people should take full advantage of the environment.	
PER 6 (Instrumental)	0.79
It is right to take advantage of the environment. Otherwise all the useful resources which we get from the environment will be wasted.	
PER 10 (Symbiotic)	0.77
We take advantage of the environment for the smooth functioning of the whole universe/creation.	
PER Factor 2: Interactiveness/Symbiosis	
PER 3 (Minimalist)	–0.44
In our country the way people show their carelessness towards the environment, it appears that they do not have any concern for environment.	
PER 7 (Spiritual)	0.55
By being close to nature, people experience a special mental poise, which shows that there is a spiritual relationship between human beings and environment.	
PER 8 (Spiritual)	0.52
Spiritual development of human beings is possible only when one is close to the environment.	
PER 11 (Symbiotic)	0.63
If we take advantage of the environment, then in return we also take measures to protect the environment.	
PER 12 (Symbiotic)	0.86
Environment will remain so long as human beings protect it.	

Cont'd...

Cont'd...

PER Factor 3: Negative Orientation	
PER 1 (Minimalist) Environment has its own place and human being has its own, and there is no significant relationship between the two.	0.77
PER 2 (Minimalist) Environment is an inanimate (lifeless) thing and human being is a living entity, so there is no question of interaction between them.	0.40
PER 5 (Instrumental) There is nothing wrong if human being has to exploit the environment completely for the welfare of humankind.	0.72
PER 7 (Spiritual) By being close to nature, people experience a special mental power, which shows that there is a spiritual relationship between human beings and environment.	–0.47

PER Factor 4: Spiritualism	
PER 2 (Minimalist) Environment is an inanimate (lifeless) thing and human being is a living entity, so there is no question of interaction between them.	–0.73
PER 8 (Spiritual) Spiritual development of human beings is possible only when one is close to the environment.	0.64
PER 9 (Spiritual) The way we worship trees and rivers, it appears that environment is a manifestation of divine and supernatural powers.	0.54

* Only those items were included in the factor structure which had at least a loading of 0.40.

As stated earlier PER view factors obtained in the present study were similar, but not identical, to those proposed by Stokols (1990). Utilitarian attitude is reminiscent of an instrumental view, and spiritualism is the equivalent of the spiritual view. Negative orientation, which implies an active negation of a PER, may resemble the minimalist view but has a negative rather than neutral implication. There is no equivalent component for negative orientation in Stokols' view. The constituent items of negative orientation showed the expected pattern of negative loadings. Similarly, the interactiveness/symbiosis factor also does not appear directly in Stokols' view, but this could possibly be part of what Stokols called the spiritual view. This factor provides further evidence of the harmonious man-nature relationship which is fostered by Indian philosophical thought.

An apparently anomalous finding was that both the utilitarian attitude and spiritualism components received the greatest degree of endorsement, although the two signify opposite philosophies with respect to PER. This may be explained on the basis of the co-existence of a traditional philosophy that accounts for a spiritual view of PER, along with economic necessity which makes subsistence dependent on the natural environment, and thereby makes Indians sensitive to its instrumental and utilitarian aspects. Additional indications of a utilitarian attitude were obtained in the reasons mentioned by the respondents considering the environment important; all of them stated that the environment is important because it provides resources. Tempered by a spiritual view of the man-environment relationship, such a utilitarian attitude may not imply an overall negative PER view.

The above piece of empirical research offers very valuable information in making it clear that people might be having all types of rituals indicating their close relationship with the environment, while day-today needs drives them toward the path, which may not be that 'positive' as it ought to be. Rather, economic necessities and struggle for survival are the issues at the bottom of the solution for environmental protection.

Looking at the above piece of research, it becomes very perspicacious that in Indian society there exists a paradoxical relationship between people and environment. Economic necessities of life have forced the people to use the environment for their benefits and survival. Unless economic needs of people's life are addressed properly, implementing schemes related to sustainable environment protection may not be very successful.

Also, if one takes these findings to the global level then there are certain imperatives. As it has become quite clear through scientific research that environmental problems and their solutions are not just country-specific, therefore a comprehensive approach is the need of the hour. Developed world has to come forward to share the cost of environmental protection for developing and poor nations.

REFERENCES

Altman, I., and Rogoff, B. (1987) "World Views in Psychology: Trait, Interactional, Organismic, and Transactional Perspectives". In D. Stokols and I. Altman (eds.), *Handbook of Environmental Psychology*, New York, Wiley, 7–40.

Merchant, C. (1992) *Radical Ecology: The Search for a Livable World.* New York, Routledge.

Ray, A. (1990) "Land Ethics' and Tagore's Vision of Ecological Harmony in Modern Environmental Policy-making for Improvement of Quality of Life", *Journal of Human Ecology*, 1, 301–303.

Sen, G. (1992) *Indigenous Vision: Peoples of India, Attitudes to the Environment*. Sage Pub., New Delhi and India International Centre, New Delhi.

Stern, P.C. and Dietz, T. (1994) "The Value Basis of Environmental Concern". *Journal of Social Issues*, 50, 65–84.

Stokols, D. (1990) "Instrumental and Spiritual Views of People-environment Relations". *American Psychologist*, 45, 641–646.

Tripathi, N. (2003) "Privacy Preference and People-environment Relationship View: An Indian Perspective", *Psychological Studies*, 48, 17–29.

Part III

Ethics, Business and Environment

EDITOR'S INTRODUCTION

Micro-level behavioural change leads to macro-level impacts on environment. We proceed from conceptual issues to attitudinal issues to business related issues. How ethical behaviour on the part of business can conserve the environment at the same time, make the business sustainable, is the issue which is considered in this section. The concept of corporate governance, corporate environmental responsibility and its practice by today's business world is elaborated in the following three chapters. This section starts with a conceptual chapter on ethics and governance which is followed by application of different social responsibility principles by the business world which is further followed by the case of India where green initiatives have been taken by a business organization and how they are related to the culture, values and ethics of that group. The linkages between philanthropy and corporate environmental strategies have also been established.

7

Environmental Ethics is the Corporate Responsibility

An Indian Perspective

ANANDA DAS GUPTA

In the age of globalization, corporations and business enterprises are no longer confined to the traditional boundaries of the nation-state. One of the key characteristics of globalization is the expanse of the market and the change in the mode of production. The centralized mode of production has given way to a highly decentralised mode of production spread across the world.

In the last twenty years, multinational corporations have played a key role in defining markets and influencing the behaviour of a large number of consumers. The rules of corporate governance have changed too. And, there has been a range of reactions to this change. On the one hand, globalization and liberalization have provided a great opportunity for corporations to be globally competitive by expanding their production-base and market share, on the other, the same situation poses a great challenge to the sustainability and viability of such mega-businesses, particularly in the context of the emerging discontent against multinational corporations in different parts of the world. Labourers, marginalized consumers, environmental activists and social activists have protested against the unprecedented predominance of multinational corporations.

As the world business environment changes, so do the requirements for success and competitiveness. Because of the forces at work, building deeper and more strategic relationships with customers, suppliers, employees, communities and other stakeholders (the corporate ecosystem) can become central to competitiveness and even survival. Building these relationships can form the foundation for a new, progressive and people-centred corporate strategy which attacks the sources—not the symptoms—of challenges facing business today. This brings us to the increased importance of corporate social responsibility (CSR).

In Western Europe, Japan, and North America, an increasing number of companies find that it makes good business sense to fully integrate the interests and needs of customers, employees, suppliers, communities, and our planet—as well as to those of shareholders—into corporate strategies. Over long a view, this approach can generate more profits and growth. Sometimes referred to as the 'stakeholder concept', it implies that management's task is to seek an optimum balance in responding to the diverse needs of the various interest groups and constituencies affected by its decisions, that is by those that have a 'stake' in the business. By including societal actors—not just financial interests—the stakeholder model assumes that enterprise has a social responsibility.

THE ENVIRONMENTAL DIMENSIONS

Worldwide, a great environmental progress has been achieved in the realm of institutional developments, international cooperation, public participation, and the emergence of private-sector action.

Over the last 25 years, eco-consciousness has been surfacing in all industrial countries, and it has proved to be powerful politics. As a result, a number of industrial countries have introduced changes that can serve as examples of 'best practices' for others.

Legal frameworks, economic instruments, environmentally-sound technologies, and cleaner production processes have been developed and applied, particularly in industrial countries.

The levels of water has gone down and air pollution in most industrial countries have declined over the past two decades, though a number of other local environmental indicators improved as well.

Due to the availability of new and better technologies, the rate of environmental degradation in developing countries (atmospheric sulphur dioxide, for example, and soot and smoke) has been slower than that experienced by industrial countries when they were at a similar stage of economic development.

The Political Dimension

Perhaps the most significant and remarkable changes over the last decades have occurred in the political arena. The number of relatively pluralistic and democratic regimes increased impressively, particularly since 1989. Good governance became a major issue to be discussed frankly on the international development agenda.

Institutional development is no longer conceived of as a process of strengthening only public institutions (which reinforced the dominance of the state and weakened public accountability) but also the private sector and non-governmental organizations.

Strengthening the role of women in sustainable development efforts is much more widely accepted and more systematically considered in practical work.

About 22 years ago the United Nations devoted a whole conference to environment. The greatest concern at the 1972 Conference on "The Human Environment" in Stockholm was the containment and prevention of industrial pollution. All facts which were available at that time were compiled and evaluated. The conference, which brought together industrialized and developing nations to delineate the "rights of the human family to a healthy and productive environment", culminated in an impressive declaration and action plan against international threats to the environment. The United Nations Environmental Programme (UNEP) was established to deal exclusively with environmental assessment and management measures.

Governance is the art of public leadership—the leadership to choose the appropriate form of political management and to create a responsible process by which authority is exercised in the management of a country's economic, ecological, and social resources. The criteria that constitute good governance include:

> legitimacy of government (degree of "democratization" and involvement of civil society organizations); accountability of political and official elements of government (media freedom, transparency of decision-making, accountability mechanisms);
>
> competence of governments to formulate policies and deliver services; and respect for human rights and rule of law (individual and group rights and security, framework for economic and social activity, participation).

A couple of years ago, the United Nations Conference on Environment and Development, which was attended by more than a hundred heads of state and thousands of political, technological, scientific and private sector delegates, also discussed themes touching on the complex, reciprocal relationships between the environment and the economy. The core areas of concern again included massive deforestation, the encroachment of desert zones, acid rain, and the damaged ozone layer of the atmosphere. Also discussed were the endemic poverty of the Third World and the excessive consumption rate of advanced nations.

BUSINESS IMPORTANCE

Companies employing a value-based business approach have found that it improves financial performance, increase employees' motivation and commitment to work, and enhance customer loyalty and corporate reputation. These and other benefits are outlined below:

Improved Financial Performance: Although it is difficult to isolate a direct causal relationship between bottom-line financial performance and a company's explicit statement of and/or implicit commitment to core values, a number of studies suggest that such a link exists

Greater Employee Commitment and Motivation: A clearly stated and demonstrated corporate commitment to certain values and goals can help employees find meaning and purpose in their work, care about what the company is doing, and link their individual efforts to those of the company as a whole. Studies show that this perception of shared values and purpose contributes to organizational performance by: (i) motivating employees, (ii) inspiring feelings of commitment, contributing to increased retention, and (iii) help attract prospective employees

Enhanced Customer Loyalty: Increasingly, consumers are factoring companies' business practices and perceived values into their purchasing decisions. Several companies that are typically associated with value-based business practices in the media and by consumers credit their commercial success, in large part, to brand loyalty among customers who support the company's values and/or mission.

Minimized Risk: Companies that institutionalize a value-based decision-making process can proactively address a broad range of legal and ethical dilemmas. As a result, these companies reduce their vulnerability to misconduct and the damage this can do to profitability, brand image, and management focus. Clearly defined values, when integrated into a company's organizational infrastructure, provide employees with the necessary tools and conceptual framework to make independent decisions in their daily operations that are consistent with the company's underlying values.

INDIAN CONTEXT

India started to develop distinctive forms of environmental laws and regulations in the 1970s. The first of India's modern environmental laws was the Water (Prevention and Control of Pollution) Act of 1974, which established the Central and State Water Pollution Control Boards; the Water Cess Act of 1977; the Air (Prevention and Control of Pollution) Act of 1981; and the Environment (Protection) Act of 1986. The latter is an umbrella legislation designed to provide a framework for Central government. The problem noticed here is not insufficient laws or pollution control boards that can control pollution, but, as the World Bank has stated, these boards have been plagued "by poor enforcement due to political interference ... whereas as with other enforcement activities in India, corruption is pervasive" (US-AEP, 1996).

Another point worth noting is that the mandate of the Central Pollution Control Board (CPCB) is to set environmental standards for all plants in India, lay down ambient standards, and coordinate the activities of the State Pollution Control Boards (SPCBs). Unfortunately, the implementation of environmental laws and their enforcement are decentralized and so is the responsibility of the SPCBs (Mani, Pargal and Huq, 1996). This is another haphazard method of addressing the issue.

In addition, pollution laws have achieved little success. The courts have been slow to respond to enforcement actions sought by state pollution boards. The boards themselves are poorly funded and charges of corruption have been regular and widespread. Large industries have achieved pollution compliance more easily than small industries (US-AEP, 1996). The reason is that they are afraid of taking risks. Lau and Srinivasan's (1997) research on identifying the driving force for better environmental performance found results that implied the current effort in environmental management is driven largely by a fear of the penalty that can be imposed by the government when environmental laws are violated. However, Cornell and Shapiro (1987) explained that a firm's value depended on the cost of explicit and implicit environmental claims. Explicit claims of the shareholders can be recognized, but the implicit claims of the firm cannot be ignored. If the firm refuses to comply with its social responsibility and quality of service, then parties to implicit contracts, like consumers or regulatory agencies, can force burdensome explicit contracts on the firm. Cornell and Shapiro's explanation applies widely to large industries, but in the case of small firms it is apparent from the literature above that this can be totally dismissed by resorting to other means.

POPULATION VERSUS ENVIRONMENTALISM

Lastly, population issues are a major growing problem in India. To sustain this growth, economic development and industrialization are the compounding factors that are an environmental burden. An approach first used by Ehrlich and Ehrlich (1990: 132–34) was to consider per capita environmental impacts on citizens. Their notion that we live in a finite world in which euphoric economic growth and population expansion would eventually exhaust the natural resources was not met without controversy. Stikker

(1992) further expanded Ehrlich and Ehrlich's approach and defined the environmental burden as:

Global Environmental Burden = Global Population × GNP Per Capita × Environmental Impact Per Unit of GNP

If one considers population, GNP, and environmental impact per unit of GNP, India, with an annual population growth of 1.91 per cent, increases every year by the size of the population of some countries in Europe. India's population doubled in the last 30 years and is expected to surpass China's population early in the 21st century (US-AEP, 1996). On an individualistic approach, if one considers individual country populations, India and China would be the countries with the highest burden. What Stikker, Ehrlich and Ehrlich suggest is that if industry is to operate within a level of global environmental quality that is not deteriorating, it must reduce its global environmental impact. Peattie's (1995: 1–15) suggestion regarding the concern generated by the Ehrlichs and Stikker concentrated on the issues of shortages of economically important resource inputs. The problems that emerged in the 1980s were not concerned with the inputs but dealt more with the environmental impact of the outputs due to indiscriminate economic growth.

Resources will be required to sustain population growth, as well as the outputs of industrialization and economic growth. Said (1997) provides two logical solutions to the Erlichs' theory, one is that humanity should breed less and the other is that we should consume less and produce lower environmental impacts in the process. Changing lifestyle and consumption habits in the long run will be effective but will not directly address environmental impact unless industries themselves take the stance and priority of protecting the environment and measures are enforced to curb population growth. The major environmental concerns in India today are poverty coupled with growing population and the side effects of enhanced industrial activities. As long as poverty remains the main stumbling block, industrialization provides hope of significantly improving the standard of living. One of the measures most talked about that might gain recognition within these industries is sustainable development. Removal of poverty and environmental protection are two sides of the same coin that is sustainable development (Dwivedi and Khator, 1995), but policy makers, governments, politicians, and industrialists have challenged many of the underlying values and assumptions of sustainability. Sustainability or sustainable development can also be described as development or progress that meets the needs of the present without compromising the ability of future generations to meet their own needs.

As regards regulatory pressure and compliance, many businesses spend more time in fighting regulations and take a less proactive, strategic approach to environmental management (Schoemaker and Schoemaker, 1995).

Johannson (1994) addresses a 'green firm' as one that does not look at regulatory or legal compliance as a first step. The ability to assure that a firm is 'in compliance' is therefore a poor tactic, and very cost-ineffective. Managers who understand

Figure 7.1 Integrative Corporate Model for Green Objectives

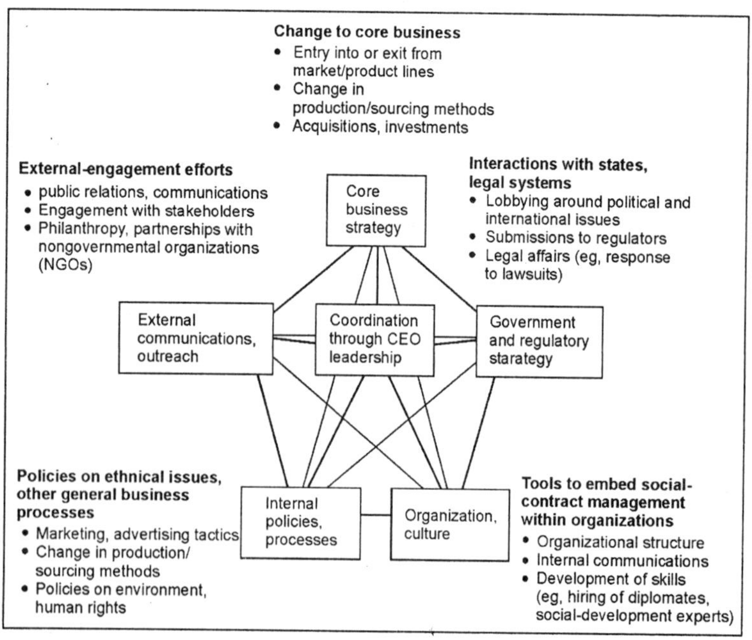

environmental laws can be counted on. In other words, regulation, compliance, and environmental laws will take care of themselves if managers adopt a sustainable vision or green objectives for industries. Here comes an integrated model for green objectives of business (Figure 7.1).

THE CHALLENGE BEFORE INDIA

Accountability is a key requirement of good governance. Not only governmental institutions but also the private sector and civil society organizations should also be accountable to the public and to their institutional stakeholders. Who is accountable to whom varies depending on whether decisions or actions taken are internal or external to an organization or institution. In general an organization or an institution is accountable to those who will be affected by its decisions or actions. Accountability cannot be enforced without transparency and the rule of law.

In the past twenty years, multinational corporations have played a key role in defining markets and influencing the behaviour of a large number of consumers. The rules of corporate governance have changed too. And there has been a range of reactions to this change. On the one hand, globalization and liberalization have provided a great opportunity for corporations to be globally competitive by expanding their production base and market share and on the other, the same situation poses a great challenge to the sustainability and viability of such mega-businesses, particularly in the context of the emerging discontent against multinational corporations in different parts of the world. Labourers, marginalised consumers, environmental activists and social activists have protested against the unprecedented predominance of multinational corporations.

There are three emerging perspectives that inform corporate social responsibility.

First, there is a business perspective that recognizes the importance of 'reputation capital' for capturing and sustaining markets. Viewing this, corporate social responsibility is basically a new business strategy to reduce investment risks and maximize profits by taking all the key stakeholders into confidence. The proponents of this perspective often include corporate social responsibility in their advertising and social marketing initiatives.

The second is the eco-social perspective. The proponents of this perspective are the new generation of corporations and the new economy entrepreneurs who created a tremendous amount of wealth in a relatively short span of time. They recognize the fact that social and environmental stability and sustainability are two important prerequisites for the sustainability of the market in the long run. They also recognize the fact that increasing poverty can lead to social and political instability. Such socio-political instability can, in turn, be detrimental to business, which operates from a variety of socio-political and cultural backgrounds.

From the eco-social perspective, corporate social responsibility is both a value and a strategy for ensuring the sustainability of business. It is a value because it stresses the fact that business and markets are essentially aimed at the well-being of society. It is a strategy because it helps to reduce social tensions and facilitate markets.

For the new generation of corporate leaders, optimization of profits is the key, rather than the maximization of profit. Hence there is a shift from accountability to shareholders to accountability to stakeholders (including employees, consumers and affected communities). There is a growing realization that long-term business success can only be achieved by companies that recognize that the economy is an "open subsystem of the earth's ecosystem, which is finite, non-growing and materially closed" (Herman E. Daily in "Sustainable Growth? No thank you" in *The Case of the Global Economy*, (Eds.) Jerry Mander, Edward Goldsmith; Sierra Book Club, 1996).

There is the third and growing perspective that shapes the new principles and practice of corporate social responsibility. This is a rights-based perspective on corporate responsibility. This perspective stresses that consumers, employees, affected communities and shareholders have a right to know about corporations and their business. Corporations are private initiatives, true, but increasingly they are becoming public institutions whose survival depends on the consumers who buy their products and shareholders who invest in their stocks. This perspective stresses accountability, transparency and social and environmental investment as the key aspects of corporate social responsibility.

Most business leaders would agree that they are managing in times of turbulence and accelerating change. They would also find a consensus about most of the trends and forces, which are challenging their traditional views of competitiveness and of the success factors for survival and profitability. These forces include the following:

Globalization of markets, consumer preferences, supplies chains and financial flows.

Some business leaders consider globalization to be a revolution, not simply a trend, since it is having momentous effects on the economies of all countries and on corporations in most sectors.

Demographic changes not only threaten the sustainability of our planet but create a mismatch between jobs and suitably-trained workers, and between present educational systems and the needs of a knowledge and information-based economy.

Environmental challenges caused by pollution and resource depletion test the sustainability of our planet earth. Business leaders are called upon to play an important role in meeting these challenges.

India is moving towards promoting Corporate Environmental Management System (CEMS) through support for the development of globalization which is already a force for great change, not simply a spectre on the horizon. Through technology, communications and economics, globalization and our increasing interconnection are inevitable. Time, distance and geography are disappearing: globalization is a reflection of that reality. In this environment, companies can and do make an important contribution to sustainable development in communities where they operate. Certain progress has already been made in some areas. But we must temper our optimism with the awareness that there remains a lot to be done, and many challenges to keep in mind as we strive to ensure that globalization is for the benefit of all people, in all countries of the world.

THE WAY FORWARD

The integration of environmental and other sustainable development considerations into everyday business is a long-term challenge for the business community, and to this end the development and use of corporate EMS tools show strong promise. It will require constant attention and commitment to respond fully in practice to the call of the Charter and similar industry codes of practice. Partnership between the public and private sectors shows great promise for the increased use of EMS and related tools in a voluntary way. Such efforts will contribute to a harmonization not just of environmental regulation and enforcement, but also of corporate policy and practice, as companies innovate and voluntarily implement and disseminate EMS wherever they do business.

As countries rely on the market system, investment and trade for the functioning of their economies, they look to the private sector as a primary source of employment creation, information, training, capacity building and education. However, if the private sector is to make its full contribution to improving their daily operations in keeping with sustainable development, an essential prerequisite is a sound policy framework, both at national and international levels, which will promote and encourage growth and development.

REFERENCES

Cornell, B. and Shapiro, A. (1987) "Corporate Stakeholders and Corporate Finance", *Financial Management*, 16, 5–14.

Dasgupta, S., Laplante, B. and Mamingi, N. (1998) "Capital Market Responses to Environmental Performance in Developing Countries" (World Bank Policy Research Working Paper # 1909), Retrieved from the World Wide Web: http://www.worldbank.org/nipr/work_paper/index.htm

Dwivedi, O.P. and Khator, R. (1995) "India's Environmental Policy, Programs and Politics". In O.P. Dwivedi and D.K. Vajpeyi (eds.), *Environmental Policies in the Third World: A Comparative Analysis* (pp. 47–70), Westport, CT, Greenwood Press.

E.N. Fitzpatrick (2000) *Ecologically Sustainable Development or Sustainable Development*, ATSE Focus,

Ehrlich, P. and Ehrlich, A. (1990) *The Population Explosion*, London, Arrow Ltd.

Friedman, M. (1970) "The Social Responsibility of Business is to Increase its Profits", Reprinted in Hoffman, W.M. and Frederick, R.E. (1995) *Business Ethics. Readings and Cases in Corporate Morality*, McGraw Hill Inc.

Gupta M. and Sharma K. (1996, third quarter) "Environmental Operations Management: An Opportunity for Improvement", *Production and Inventory Management Journal*, 40–46.

Hart, S. (1997, January/February) "Beyond Greening: Strategies for a Sustainable World", *Harvard Business Review*, 67–76.

http://www.atse.org.au/publications/focus/focus-fitzpatrick2.htm

Johannson, L. (1994) "Profits From Green Ventures", *The Management Accounting Magazine* (CMA), 68(7), 17–21.

Lau, R. and Srinivasan, R. (1997) "Strategic Issues of Environmental Management", *South Dakota Business Review;* 56(2), 1, 4.

Mani, M., Pargal, S. and Huq, M. (1996) "Does Environmental Regulation Matter? Determinants of the Location of New Manufacturing Plants in India in 1994" (World Bank Policy Research Working Paper # 1718) Retrieved from the World Wide Web: http://www.worldbank.org/ nipr/work_paper/index.htm

Nyati, K. (1988) *Problems of Pollution and its Control in Small Scale Industries*, New Delhi, India, Friedrich Elbert Foundation.

Pargal S., Mani, M. and Huq, M. (1997) "Inspections and Emissions in India: Puzzling Survey Evidence on Industrial Water Pollution" (World Bank Policy Research Working Paper # 1810) Retrieved from the World Wide Web: http://www.worldbank.org/nipr/work_paper/ index.htm

Parthasarathy, S. (1996) "Whither Small Scale Industry?" In *The Hindu Survey of Indian Industry*, pp. 14–26, Madras: Kasturi and Sons, National Press.

Peattie, K. (1995) *Environmental Marketing Management: Meeting the Green Challenge*, London, Pitman.

Porter, M. and Linde, C. (1995, September/October) "Green and Competitive: Ending the Stalemate", *Harvard Business Review*, 120–134.

Prof. Dr Klaus M. Leisinger's lecture given at the 9th Annual Conference of *The European Business Ethics Network* "Working Across Cultures" Frankfurt, September, 18–20, 1996.

Quote from the Executive Summary of United Nations Environment Programme: *Global Environment Outlook*. Oxford University Press, New York, 1997, p. 2.

Rajendran, S. (1989) "Institutional Support to SSIs". In *The Hindu Survey of Indian Industry*, pp. 279–283, Madras, India, Kasturi and Sons.

Roberts, Michael (1995) "Indian Public Demands Industry Responsibility", *Chemical Week*, 157(1), 89.

Said, David (1997) *Green Australia: Mapping the Market*, St. Leonards, N.S.W., Australia, Prospect Publishing.

Schoemaker, P. and Schoemaker, J. (1995, spring) "Estimating Environmental Liability: Quantifying the Unknown", *California Management Review*, 29–61.

Shaman, D. (1996) "India's Pollution Regulatory Structure and Background". USA: The World Bank. Retrieved from the World Wide Web: http://www.worldbank.org/nipr/india/india-back.htm

SIDBI Report on Small Scale Industries Sector, (1999) Lucknow, India: Small Industries Development Bank of India.

Stikker, A. (1992) "Sustainability and Business Management", *Business Strategy and the Environment*, 1(pt. 3), 1–8.

Sustainability Website: www.sustainability.com

The World Commission on Environment and Development: Our Common Future, Oxford University Press, 1987, p. 8.

Tiwary, J. and Singh, B. (1990) "Industrial Development and its Changing Pattern in India During the Plan Period". In *The Indian Economic Association*, conference volume, 73rd Annual Conference Muzaffarpur, pp. 46–54, n.p.

UNDP: *Human Development Report 1997*, Oxford University Press, New York, 1997.

UNICEF: *The State of the World's Children* 1998, New York, 1998.

United Nations Environment Programme: *Global Environment Outlook*, Oxford University Press, New York, 1997, p. 2.

United States-Asia Environmental Partnership. (1996) *Industry and Environment in Asia: US-Asia Environmental Partnership*, Retrieved from the World Wide Web: http://www.usaep.org

Vedantam, V. (2000) "India's Billionth", *The Christian Century*, 117(26), 945.

Welford, R. and Bhargava, S. (1996) "Corporate Strategy and the Environment: The Theory", In Richard Welford (ed.), *Corporate Environmental Management: Systems and Strategies*, pp. 13–32, London, Earthscan.

WHO: The World Health Report 1998. *Life in the 21st century, A vision for All*, Geneva, 1998.

World Bank: *Governance and Development*. Washington, D.C. 1992, p. 9 f.

World Business Council for Sustainable Development Report, *Corporate Social Responsibility*.

World Commission on Environment and Development (1987) *Our Common Future*.

World Development Report (1997) *The State in a Changing World*, Oxford University Press, New York, 1997.

Worldwatch Institute: *Vital Signs* (1998) W.W. Norton and Company, New York, 1998.

8

Environmental Initiatives and Corporate Practices

The Case of Hind Lamps Ltd. of the Bajaj Group

PARUL RISHI, PARAG DUBEY AND SWATI MOGHE

In the era of globalization and cut-throat competition, it is not possible to spend on anything which does not make business sense. The idea of corporate social responsibility (CSR) is not something new for the corporate world. For a last decade or so with the wake of transnational and multinational corporations on the Indian economic horizon, the established ways of doing business in India have come increasingly under global influence. The norms of production, corporate governance, labour and corporate practices are changing, and more and more organizations are learning to deliver to their stakeholders more than what is expected by the law. Even though the process is slow and the adoption of corporate social responsibility as a business strategy is not widespread, the change is indeed taking place. There has been a shift as far as the understanding and implementation of CSR is concerned. Corporate social responsibility, rather than being a philanthropic activity, is being looked by modern managers as an integral part of business proposition itself (Arora, Garg and Vaida, 2007).

Gorte (2005) has mentioned that social responsibility has long been a focus of attention and interest of business managers and academicians. Many management professionals view social responsibilities as a paternalistic pursuit where the managers assess what the society requires and subsequently undertake actions to accommodate the said needs (Crossley, 1999; Trainer, 2005). In reality, the operations of an enterprise need to be driven by a larger set of its objectives, which are today being defined under the term Corporate Social Responsibility (CSR). It is essentially a concept whereby companies integrate social and environmental concerns with their business operations and in their interaction with stakeholders on a voluntary basis. This means not only fulfiling legal expectations, but also going beyond compliance and investing in human capital, the environment and relations with stakeholders.

CULTURE AND VALUES FOR CORPORATE SOCIAL RESPONSIBILITY

To gain a better understanding of how corporate or industry cultures can change overtime, it is important to know the business in which they operate. Culture of any corporate body plays an important role in adopting and developing the feeling of social responsibility among their members. It is essentially concerned with the ethics, attitudes and influences, the behaviour of an individual, groups and process. A popular conceptualization of organization culture has been through the shared reflection of organizational reality to a particular group or organization (Mishra and Gupta, 2007). The culture eventually evolves in a particular organization. It is a complex outcome of external pressure, internal potentials, and responses to critical events and chance factors that could not be predicted from knowledge of either the environment or the members. Imposing new culture would probably be met with resistance, as it is difficult to change the core values (Emery and Oertel, 2006). Culture and values of any organization are highly intertwined. Positive values can create a competitive advantage for organizations. They can be a unifying force which brings efforts to bear on highly desirable outcome. A positive and clear ethical climate helps organization members to make good choices when facing ethical dilemmas (Barney, 1986). A strong culture with unique values can, however, be a double-edged sword to provide a strong corporate identity, enhance collective commitment, provide a stable social system, and reduce the need for formal and bureaucratic controls.

Figure 8.1 Open System Model for Hind Lamps

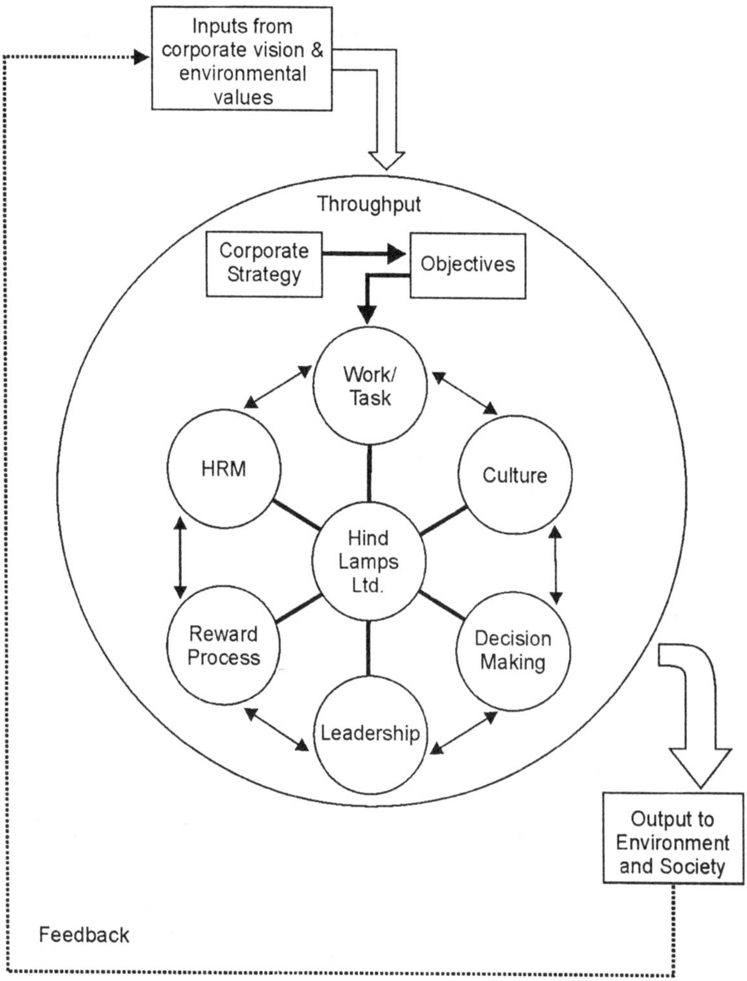

Some of the factors that may be emphasized in different ethical climates of organizations are personal self-interest, operating efficiency, company profit, team interest, social responsibility and personal morality. These are (ethics, rituals, symbols) powerful aspects that can be important in establishing and maintaining a work culture and know-how to act and expect others to act in various circumstances. They provide a common orientation to decision-making and modes of action that can facilitate performance (Schermerhorn, Osborn and Hunt, 2004).

Culture and its underlying values are the driving force of any organization (Purang and Sharma, 2007). The extent to which values are consciously incorporated into CSR decision-making processes indicate an organization's value-attunement or value-neglect, a concept developed by (Swanson, 1995). This sub-dimension recognizes that CSR is a value-laden process, that is, "an ideal type of responsiveness that involves the selection, retention, and enactment of organizational values that are consistent with constructive social impacts" (Smith et al., 2001). Charles Thuard (2005) believe that an organization is able to fulfil its social responsibility by regularly engaging in CSR activities, which enhances the value system of the organization. This, in turn, helps in developing its brand, customer and employee loyalty. Besides, employees feel a sense of pride for belonging to the organization which is committed to the society.

The model of an open system approach of organization indicates that every organization exists in an environment from where it derives its inputs (Figure 8.1). In regard to CSR, Hind Lamp derives its inputs from its corporate vision which is marked with philanthropy coupled with environmental values. These values act as a major guiding force in the transformation or throughput process which starts from corporate strategy and its specific objectives in relation to human resource (HR) policy, organization culture, reward process, decision making and leadership style. If any of these organizational components are not designed and/or functioning in line with the guiding corporate mission/environmental values, it will affect the output to environment and society and the organization will not be able to get the objective of CSR activities fulfilled in the real sense of term. Therefore, there is a need for continuous feedback from the internal and external environment of organization and society to refresh the corporate vision and values and resulting throughput process.

Hence, the concept of corporate social responsibility came up and many organizations took the initiative for environmental conservation as the part of their CSR activities. These environmental issues have also become an important factor in the decision-making process of companies around the world. Through voluntary commitment to CSR, companies hope to send a positive signal to their various stakeholders (employees, shareholders, investors, consumers, regulators and NGOs), and in so doing make an investment in their future and help to increase profitability.[1]

LEADERSHIP WITH AN ENVIRONMENTAL VISION

A close analysis of high performing corporate bodies shows that a leader must address three central tasks: reinforcing environmental values, building community culture and developing vision with environmental values as a prerequisite to success of the corporate for its social and environmental responsibility. Many thinkers argue that leadership requires power as a prerequisite, but effective leadership is not merely the ruthless application of power to achieve organizational goals or personal satisfaction.

The traditional thoughts of leadership have not much to say about the environmental part of corporate social responsibility. Therefore, the leadership based on the traditional sources and methods are not completely equipped to solve the social and environmental challenges effectively, and therefore leaders may require special efforts and attention on the environmental challenges which they are facing in today's life. The corporate that formerly viewed environmental regulations as constraints are increasingly striving to view them as sources of competitive advantage and the 'environmental managers' are now beginning to replace as 'environmental lawyers' in corporate culture (Gordon and Berry, 1961). In the context of corporate social responsibility, there is a need for leaders who can visualize business more than a profit earning activity. A deep compassion for society, environment and people clubbed with the growth and production oriented motive for organization, is a prerequisite for leaders who visualize his organization holding corporate social responsibility. There seems to be a change towards more participatory and open model of leadership rather than the traditional leadership model. Other areas of consideration in CSR include the basic values of the organization which influence the behaviour and actions of the leaders and define their character and personality. Companies may be motivated to carry out CSR for moral reasons. Voluntary CSR goes beyond the usual economic confirms of social responsibility in contributing to the common good at the expense of the business (Lantos, 2001). Many multinational companies have agreed with the ethical theories on social responsibility and have developed codes of conduct for issues such as child labour, gender or racial discrimination, promotion of human rights etc. (Mittal, 2007). The external world perceives and responds to an organization based on the way it goes about conducting its business. The world of business even today is going through a crisis of trust, where the credibility of business leaders and the very legitimacy of business organizations is being questioned. The scrutiny from the government, the media, analysts, investors and action groups is becoming more stringent everyday. People and society are looking for the soul of the corporation and its ethics. There are several forces that drive business towards a quest for values. For example, it is an enlightened founder or the CEO who sets the tone and drives ethics deep through the organization. This leads to the premise that values and business objectives are not mutually exclusive. Doing business with a strong sense of ethics is a win-win game for all. The values and behaviour of transformational leaders can influence strategic choices like CSR, through their impact on their field of vision of organizational members, their perception and interpretation of their decision and strategic decision-making (Waldman et al., 2004). It may be that for leaders to get so wrapped up in their organization's pursuit of a social cause as to personal ethics have compromised in the process. It hardly makes sense for corporate social responsibility to be pursued at the expense of responsible and ethical action towards employees. Somewhere in between such conflicting forces, leaders have to make the difference.

STRATEGIES FOR ENVIRONMENTAL AND SOCIAL RESPONSIVENESS

The concept of corporate social and environmental responsiveness is the ability of business to interact positively and humanely with its social and environmental surroundings. Their interaction with social and environmental forces and understanding of trends help them to formulate public policies that integrate ethics into their decisions and actions.

A careful planning for long-term effects of formulated decisions and policies make a company's management socially responsible. Social responsiveness permeates both formulation and implementation of a corporate strategy. The strategies of a corporate involve basic decision about mission, purpose and reason for individuals and organizations. The decisions are value-laden decisions, affected by the values and ethics of management.

Community demands for improved corporate environmental performance have more recently been broadened to include expectations for greater involvement in the overall development of local communities and for stricter ethical standards governing corporate behaviour. These new expectations reflect the growing recognition that no segment of society can take total responsibility for human progress, and that corporations operating in free market environments can make large contribution to social and economic development in partnership with government and other outside agencies.

The interest of various stakeholders and the web of social and environmental issues are a vital dimension of corporate responsiveness. Another area of corporate performance that has begun to capture the attention of professionals is the environmental responsiveness or the green initiatives, which increases the environmental and societal well-being. These green initiatives are a set of responsibilities that contribute directly to an organization's management profile and are sometimes also linked with 'corporate social responsibility'. A corporate cannot select its social actions in a vacuum. It needs stakeholders' inputs, dialogue with external forces to ensure a positive impact on all the stakeholders' relationship.

Environmental initiatives and social performance matter to stakeholders, including diversity in the workforce, innovation and risk management, pollution prevention, and community development. Corporations are motivated to involve stakeholders in their decision-making and to address social challenges, because today they are increasingly aware of the importance and impact of corporate decisions upon society and the environment. The growing environment and social concerns of stakeholders present a unique opportunity for companies to give better leverage their green initiative capabilities. This will enable the corporate to measure and manage the contribution of environmental responsiveness and social performance in market and customer value (GEMI, 2004).

The social and environmental initiatives contribute to the corporate and their stakeholders in a variety of ways—not only to 'direct' benefits like risk reduction and profitability improvements, but also through 'indirect' benefits like brand equity, human capital and strategy execution. These challenges faced by the world community in the next decade are to work towards a sustainable future for the planet. Internationally business organizations and public institutions have a vital part to play, and people in all walks need to review their own styles and make their adjustments to ring in a future that will be sustainable. Industries need to prove that they are capable of operating in ways that are sustainable. This will be accomplished by treating social and environmental issues as an important strategic decision variable.

SUSTAINABILITY AND ORGANIZATIONAL STRATEGY

Social, environmental and economic sustainability has become a focus of many CSR efforts. Sustainability was originally viewed in terms of preserving the earth's resources. In 1987, the World Commission on Environment and Development published a landmark action plan for environmental sustainability. The Commission, named after former Norwegian Prime Minister Gro Harlen Brundlandt, defined sustainability as "meeting the needs of the present without compromising the ability of future generations to meet their needs".

The concept of sustainable development received further impetus from the United Nations Conference on Environment and

Development in Rio de Janeiro in 1992 (the 'Earth Summit'), when countries of the world agreed to work together to address the widely recognized environmental problems which the earth was experiencing. National governments were expected to respond locally. A company has to ensure that its operations are sustainable in regard to its economic, social and environmental performance. This stakeholder-based definition suggests that Triple Bottom Line (TBL) is an appropriate approach to measure organizational sustainability performance. In short, introducing sustainability concepts with the TBL measurement system means that the organization must think about its impact not only on the current society but also on the future generations, so that whatever they do are being pursued and are able to be sustained into the future.

For measuring the economic, social and environmental (TBL) performance of any organization many reporting systems had emerged. However, only two of these reporting systems have achieved more support than the others, the 'sustainability framework', representing the work of an international consulting firm, while the 'Global Reporting Initiative' (GRI), a United Nations affiliate, whose development is supported by a large number of global corporations. GRI recommends specific information related to environmental, social and economic performance which is closely related to TBL measurement concept.

Organizations' reactions to the issue of sustainability have showed their reactions to the environmental issues. Many organization do not distinguish between 'environment' and 'sustainability'. Organizations always try to highlight their environmental measurements rather than highlighting their social and economic measurements because nowadays preserving environment is the major concern for the society. Strategically, for many organizations, sustainability has come to mean 'consistent levels of economic growth' for the organization, or economic sustainability (Bansal, 2002), rather than the holistic view of 'sustainability'. Ehrenfeld (2005) argues that the corporate 'sustainability' reports that organizations are now beginning to produce do not, in any way, come close to reporting the requirements of true sustainability for society. Sustainability therefore is likely to have huge impacts on organization strategy. Strategically, organizations can see sustainability as a pressure to comply with, a cost to be incurred or an opportunity for competitive advantage (Hubbard, 2006).

METHODOLOGY

Hind Lamps Ltd. of Bajaj Group

The study was conducted in Hind Lamps Ltd. of the Bajaj Groups about 70 kms from Agra at Kanpur Rroad in the state of Uttar Pradesh, India. In year 1951, Bajaj Electricals took a step forward and entered into collaboration with N.V. Philips Geoeilamenpenfabrieken of Holland and three leading British electrical firms, Associated Electricals Industries Ltd., Crompton Parkinson Ltd. and GEC Ltd. to form Hind lamps. It was nearly the beginning of the electrical equipment industry.

The culture that Bajaj's have nurtured in their organization, provides excellent motivation to every individual to put his best performance in an enthusiastic and charged up environment that sans any fear of failure. They have a strong belief that given the right kind of inspiration, excitement and motivation that reaches the hearts and souls of people, and ordinary people can perform extraordinarily.

The Bajaj Group have the history of ethics based Gandhian leadership. They have a strong belief that "our wealth should be utilized for improving the quality of the society and the nation. We should never fail to spend at least a portion of our income on the social, economic and educational development of the provenance where we live" says Jamnalal Bajaj. Hind Lamps Ltd. of Bajaj Groups, follows the footsteps of their veteran leaders and shows their commitment for environment and society at large through a foundation named 'friends of nature'. Under this organization, they are undertaking various environmental activities in the nearby villages of their unit which are being heavily polluted by the local glass industry. Besides, Hind Lamps, as an institution, is also playing a leading role in facilitating various CSR activities under the leadership of various levels from CMD to executive director and executive functionaries of 'friends of environment'.

Bajaj Electricals has undertaken social well-being as volunteer practices almost from its inception; long before the term social responsibility was coined. The company's earliest initiative in community development and philanthropic activities was traced by establishing the Jamnalal Bajaj Foundation (JBF) on 4th November 1977. The foundation's basic objective was working for betterment of the community living in and around the villages near Bajaj Group factories. To make significant efforts to minimize continuous degradation of the environment, deterioration of physical and mental health, global warming, floods, drought and other man made disasters—a society 'Friends of Environment (FEN)' registered on 24th September 2004 under the umbrella of Bajaj Electricals Ltd. It has its headquarters at Shikohabad in Uttar Pradesh with 22 working groups *(Families)* spread all over India.

TOOLS AND TECHNIQUES

A mix of qualitative and quantitative tools and techniques were used to study the organization culture and CSR values of Hind Lamps from various respondents. CSR and organization culture rating scale i.e. motivational environment, equal treatment, commitment towards society and environment, environmentally safe standards, open communication, creativity and innovation, human resource practices, capacity building, decentralization, conflict management, job satisfaction, leadership style schedule, i.e. problem solving: (a) gives personal involvement in problem solving when and where we require, (b) gets impatient as and when problem arise (lose balance, gets angry), (c) encourages staff and beneficiaries to solve their problems and learn from them. Likewise, the adapted version of the schedule for executive managers was designed and feedback of the executive managers were also taken as their self-assessment, i.e. decision making: (a) "I take all the decisions alone, does not consult others, (b) I take decision by consulting a few closer to me and ignore other, (c) I take decision by consulting as many as possible, gives every one opportunity to participate in decision making." Key informants' interviews, personal interviews of the beneficiaries, focus groups discussion, documentation. secondary information were collected through the annual report of Bajaj Hind Lamps, newsletter of *Paryavaran Mitra* and booklets of JBF.

ENVIRONMENTAL INITIATIVES

Documentation of green initiatives and rural development activities of Friends of Environment (FEN) and Jamnalal Bajaj Foundation (JBF) was done through visiting the concerned villages where these activities were carried out. The details of activities carried out under green initiatives are as under:

Hind Lamps Ltd. was showing their responsiveness towards community and society by generating awareness among the community to save environment through various green initiative activities, and in line with it, they are implementing various rural development, and community welfare activities. Besides doing all the developmental and welfare activities, FEN has first started campaign for creating awareness among the staff of Hind Lamps and the inhabitants of nearby villages to keep their surrounding green and clean. Huge wall writings in various public and religious places were undertaken to urge people to 'Grow more trees' 'Protect the Environment and keep it hygienically clean'. Business promotional materials including array of items like posters, handbills, stickers, pamphlets and banners were specially created and designed with the message and to communicate a number of things that people can do to prevent air, water, noise and land pollution with a logo of *Friends of Environment* (Plate 8.1).

Plate 8.1
Business Promotional Material

Every year different environment related days like World Forest Day, World Environment Day, and World Water Day were celebrated with great enthusiasm. Additional steps for creating awareness were including organizing workshops, seminars, rallies, quizzes and painting competition on topics like 'Save trees and forest', 'Saving a drop of water is equal to saving a drop of blood', 'Avoid water pollution' etc. Anti-Pollution rallies were also organized involving the staff of Hind Lamps, FEN and JBF with slogans like 'remove pollution, and be healthy' etc.

Friends of Environment was involved in the plantation activities in Hind Lamp campus as well as the surrounding villages. A group of self-motivated volunteers of Friends of Environment under the leadership of Assistant General Manager, Hind Lamps, set many activities like cycle rally in the villages of Shikohabad to popularize the importance of growing trees. Hind Lamps had a privileged Chandi Prasad Uddayan, a plantation site planted with various shade providing tree saplings along with the staff of Hind Lamps, and now these saplings had grown into trees. This was an initiative forward to make the campus green and clean by involving the staff of Hind Lamps to promote a positive attitude towards saving the environment. On World Anti-desertification Day in year 2007, two thousand saplings of various shed providing trees were planted in Hind Lamp campus.

The use of bio-fertilizer as a substitute of chemical fertilizers was another important activity undertaken by FEN. Through many meetings with the farmers' at the local village leader's (*sarpanch*) house or any central place in the villages of Shikohabad to educate the farmers about the ill effects of using pesticides and chemicals on crops (Plate 8.2). Afterwards they were informed about vermicompost and organic farming methods. A continuous effort was undertaken in this direction by organizing awareness camps regarding the benefits

Plate 8.2 Bio-Fertilizer Production

of organic farming. As a result, many farmers started preparing vermi-compost for their own use and have thought of preparing it for sale also.

COMMUNITY DEVELOPMENT INITIATIVES

The community development activities of FEN revolve around the underprivileged community that lives in the immediate vicinity of the Hind Lamps factory. The range of these activities begin with extending education through schools for child labour and goes up to cover vocational guidance and supporting employment-oriented and income-generating projects like manufacturing of bio-fertilizers, women empowerment through self-help groups, and establishment of biogas to reduce the side effects of using firewood on the health of rural women and on the environment as a whole.

Biogas is the primary source which falls in non-conventional energy category to reduce the dependency on firewood and thereby reducing the pressure on surrounding forest areas. It is already established that bio-gas is a smokeless pollution free energy source and easy availability of cowdung as an input material in villages made it reasonable alternative to establish as a major non-conventional energy source in the villages surrounding Hind Lamps. Yarn production centre was established to empower the rural women under JBF, where facility of trainer for use of the spinning wheel for producing yarn was provided. Self-Help Groups (SHGs) under JBF were formed as a concrete step for women empowerment.

Education is imparted not only to children of Hind Lamps employees but also more importantly to children from rural areas who do not have access to any form of information or education. Under JBF many educational centres were set up for the betterment of the rural mass. Schools for child labour, play schools (*Balwarees*) for rural kids, and adult education centres to make the rural adults functionally literate were also operating successfully. Shikohabad and Firozabad district are famous for their glass articles manufacturing making use of child labour very common site. To eradicate this social ill of child labour from society, government had started a national level rehabilitation project for child labours. Many adult education centres were established to encourage the rural mass (especially women) towards functional education.

JBF had joined hands with government departments for Total Sanitation Campaign (TSC) and has constructed *sulabh* toilets. This initiative was proved quite successful in creating awareness among villagers about hygiene. To get people have pure and clean drinking water, JFB has undertaken new steps in collaboration with CAPART, New Delhi, for providing hand pumps. Health awareness camps were also organized in different villages at different time intervals to educate the villagers.

CORPORATE STRATEGIES FOR ENVIRONMENT

There has been a tremendous amount of exposure of environmental subject in national and local media in recent times and most people including stakeholders of Hind Lamps Limited are now aware of these issues. Hind Lamps have shown their social and environmental responsiveness by continuously upgrading their corporate strategy and coming in line with ISO 14001 certification for maintaining high environmental standards of production and sales. It is further reflected by the discussion with CMD, Bajaj Electricals and Executive Director of Hind Lamps. The discussion clearly indicated that they have a long history and attitude towards philanthropic activities coupled with environmental initiatives at the business level. This is further justified by their charitable assets of \$153 million, out of the total \$1,333 million of business assets which are about 10.3 per cent of their total assets. Moreover, they have spent \$2.15 million on philanthropic activities in the year 2005–06.

The discussion further reflected that they have a long-term strategic planning for the development of environment friendly technology which will consequently encourage green products, keeping in mind future challenges and opportunities. The same was also reflected during deliberations with the Executive Director, Hind Lamps, in terms of concrete action plans for marketing, production, human resource and finance of the company.

As the situation suggests, one key decision that Hind Lamps managers must take about each major environmental problem they face is whether to lead or lag behind their competitors on environmental issues. In some cases, they may be interested to pursue an environmental strategy in which it gets well in front of regulations or public opinion and in other cases they may be served by moving in lock step with industry or by reacting only in response to external pressure. If Hind Lamps lag, they may find themselves on the receiving end of unfavourable regulations, but if they take initiatives and lead, their action could increase near their production cost and leave the company vulnerable to its competitors.

In view of the above, it is very relevant to mention that Hind Lamps is an ISO 14001 certified organization which ensures its commitment for meeting the environment standards while designing and manufacturing the products and ensuring safe disposal of hazardous waste. There is a well-laid "Environmental Policy" which was made public on 15th May 2006. The policy states their commitment for the cause of environment and ensures to generate awareness in the staff at all the levels down the line for its real translation into action, but the findings indicated that such awareness is yet to take its shape as the policy is just one year old and, more awareness generating efforts at the organizational level are required in this regard.

There is a need to further strengthen environmental focus, as a part of long-term strategy to ensure a high level of environmental commitment from staff of Hind Lamps. It can be done by explicitly articulating respect for 'sustainable growth', i.e, developing products that minimize environmental impact while focusing on environmental improvement as well as the profitable for the organization. It was well documented in the interview with ED, that their environmental initiatives have given them profits, besides building the corporate green image by using propane gas as a cheaper substitute of LPG.

The basic concept of corporate social and environmental responsibility focuses on the triple bottom line principle, 'economic, social and environmental', development with the aim to accomplish social and environmental development while achieving business success. This information obtained from Hind Lamps on all the three parameters of this approach is as under:

- *Economic:* The cost saving process by minimizing the wastage of raw material and recycle of waste glass products for manufacturing had reduced the total production cost. Hind Lamps has a major future strategy to introduce energy saving products (CFLs) in their premises like other branches of Bajaj Electricals Ltd., which will be five times energy efficient than the traditional yellow incandescent bulbs. At corporate level Bajaj Electricals is earning 50 per cent of the revenue from export of CFL bulbs. Swift to use cheaper environment friendly propane gas is also a part of their cost saving mechanism.

- *Social:* The multi-stakeholder participatory endeavour for environment protection, community development and social welfare activities allow all the stakeholders to participate and understand the mission and vision of the company. This helps Hind Lamps to establish healthy relationship with all stakeholders on a common ground.
- *Environmental:* The shift in manufacturing fuel as propane gas instead of LPG is more environment friendly, as the final outcomes of propane gas, SOx and NOx are not the products which create air pollution and are not responsible for ozone layer depletion or climate change. Switched over to water-based coating for tube lights instead of lead based

Figure 8.2 Triple Bottom Line Components and Their Indicators

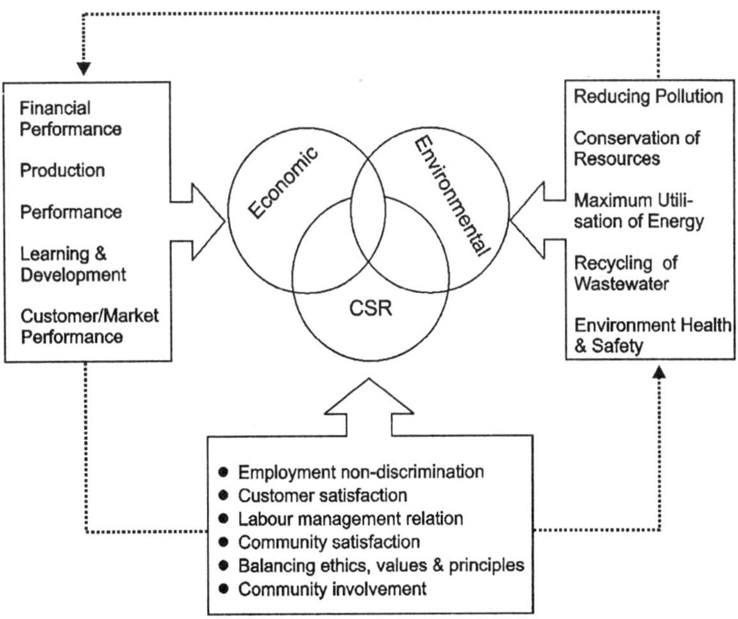

coating which was a major reason for soil pollution. Planting of two thousand sapling of different species inside the campus is a major initiative in this regard.

The 'Green Initiatives' of Bajaj Hind Lamps is working on the triple bottom line concept (Figure 8.2). They are doing environmental development by using propane gas for less negative effect on environment and making the campus green to reduce pollution. They are developing the nearby community areas and doing social welfare activities by involving all their stakeholders, as a part of social development as their social responsibility towards society. Overall, by the environmental and social development they are in line with economic success by reducing total production cost and earning revenue from energy efficient strategies at corporate level.

However, using Carroll's CSR pyramid (1991) as a framework to analyze the case, Hind Lamps can be largely placed under the 'Ethical Responsibility Model', as Bajaj Group was highly influenced by Gandhian thoughts since their inception and it is still reflecting the social welfare and environmental activities. Commencement of Jamnalal Bajaj Foundation and Friend of Environment for green initiatives, are a part of their philanthropic actions, which may not have any direct relationship with their business profits, but show their concern for the stakeholders. Formation of these social foundations at all their branches in India, may not have the involvement of all the stakeholders, i.e. dealers, distributors and customers but allowed them in developing a strong common ground between all levels of staff to achieve consistency of understanding about their social and environmental values.

ORGANIZATIONAL ANALYSIS FOR CSR

The analysis of organizational culture and managerial values for green initiatives of Hind Lamps Ltd. brings out some important issues. They are outlined as under.

Cultural Remix and Opportunities for Change

The organization culture of Hind Lamps Ltd. has been positively changed during the span of last two years. It is primarily due to change in the top leadership at organizational level. The Executive Director (ED) had a vision for growth and development of the organization, which was blended with his vast experience of working in the lightening industry since a few decades. Owing to his expert guidance and humanized approach to management, the organization which was shut down for eight months due to various technical as well as HR related crucial issues and was running losses, lately picked up momentum. As organizational health of Hind Lamps Ltd. was not adequate since long, the personal interview with ED revealed his major corporate strategies for the organization:

- Optimal utilization of capacity and resources.
- Humanized approach towards employees' management.
- Trust building mechanisms.

However, the roots of local culture and politics were still reflected in the organizational culture, leading to various HR and administrative related issues, yet to be solved. It was also noted that frequent changes in the administrative polices of the company at higher level coupled with lack of industrial culture primarily owing to excess number of local employees besides other transient reasons, has led to some administrative challenges. In view of that a need for the employees to develop healthy interpersonal relations was felt by top managerial staff. It indicated that due to the conflict avoiding strategies of Hind Lamps Ltd. management due attention was not paid to the HR issues.

HR Challenges

After analyzing the results of content analysis, it was observed that HR issue was the main factor hindering the growth and development of the organization as perceived by the respondents (Table 8.1).

Table 8.1 Content Analysis of Perceived Developmental Challenges to Hind Lamps Ltd.

Challenges in development of HL		Respondents' per cent
Production Issues	• Underutilization of plant machineries	32%
	• Non-availability of cheaper natural gas instead of LPG	48%
	• Outdated production system, still focusing to old products like—GLS bulbs and tubes	72%
	• High production cost	20%
Marketing Issues	• Poor marketing and sale	60%
	• Unfavourable market condition	8%
Financial Issues	• Limited financial resources	12%
Administration Issues	• Lack of industrial culture	4%
	• Unstable policies at higher level	28%
	• Withdrawal of Philips from Hind brand	16%
Training and Development Issues	• Lack of efficiency in man power	8%
	• Staff is not educated, motivated, concerned, and trained	16%
Human Resource Issues	• Indiscipline	20%
	• Union activity	24%
	• Strikes and lockouts	8%
	• Low morale due to low salary	48%
	• Local politics	4%
	• Absenteeism	28%
	• Lack of team work	4%
	• Discrimination between old and new staff	4%

Results regarding equal treatment in terms of caste, administrative hierarchy and economic status also supplemented the above findings. Middle level staff and FEN staff were not much influenced by casteism and hierarchy related discrimination as against senior-level managers and foremen. A similar uncertainty was existing in regard to open communication. Communication channels were found to be not so open as perceived by senior managers, FEN staff and even foremen, to some extent except just middle-level staff. There was the need for more decentralization of power and authorities at organization level. To overcome the HR issue, transparency should be maintained in the communication system and motivation of employees should be maintained for team work, without any caste or hierarchal discrimination. Very limited opportunities were available for people to have trainings in new technologies and general behavioural issues which were limiting the possibility for creativity and innovation in the organization for further growth and development. It was also observed that if proper training opportunities were imported to the employees, it will enhance their competencies and increase their work efficiency, development.

The above analysis indicated that the thought process of Hind Lamps is yet to effectively integrate their CSR values and vision, though they are in the process of providing outputs to society and environmental activities by spending substantial resources. In view of the above a little effort in regard to reviewing the organizational system and its redesigning keeping in mind the CSR values, is desirable on the part of Hind Lamps to upgrade their CSR image and its results to benefit the organization, society and its environment.

To meet these challenges, the major corporate strategy was humanized approach to management with the concern for employees' problems and treating them as a family. Besides, a non-coercive approach to discipline and nurturing task leadership style was also adopted as a step towards solving the HR issues. Though Hind Lamps was found to be thoroughly committed to social development and environmental conservation through its foundations like JBF and FEN, its policy regarding environmentally safe standards of working was not reaching its employees down the line. Production and marketing issues were indicated as the next challenge for Hind Lamps Ltd. as they were interlinked with each other.

Developmental Challenges for FEN

It was observed that FEN was actively involved in rendering the environment, green reducing pollution, encouraging the young farmers to use bio-fertilizers, providing training to empower rural women and creating awareness among the people of the surrounding rural areas. As far as the level of satisfaction of the respondents regarding the activities, they were found to be partially satisfied with the current activities of the FEN, may be because the organization is at the nascent stage and needs time to grow and develop. In this regard the case of Cotton Textile Manufacturing Company of India is worth mentioning. It highlights the point that in the days of competition when activities like CSR are considered to be undesirable but unavoidable at the same time it would be more sensible for the managers to identify the areas where the social initiatives can be undertaken and how those initiatives serve the business purpose (Arora, Garg and Vaida, 2007).

Any new initiative is smashed with a hammer, not because it is inappropriate or undesirable, but just because it is *new* and it has to encounter the inertia of organization. FEN also experienced the same in its early stage of group development and is still experiencing it to some extent. Discussions with organizational members revealed that their perceptions about FEN were not very positive and they were considering it as a sort of extra-activity from which they consciously wanted to dissociate. This can be interpreted in terms of the theory of group development (Tuchman, 1965). This theory indicates that every organization starts its life at the forming stage which is followed by the storming stage when all the conflicts and issues related to formation of organization vitiate the scenario. Gradually, the storm is quelled and over, and organizational members commonly decide certain norms of operation at the norming stage which culminates into the stage of performing. At the stage of group development, if the leader is not playing an adequate role, the group or organization carries the risk of adjourning (Figure 8.3).

In regard to current study, FEN has overcome the forming stage and may be struggling to overcome the storming stage to arrive at the norming and performing stages. As the initiative taken to form an organization Friends of Environment (FEN) for environmental conservation and social development was taken by the CMD of Bajaj Electricals Ltd. in Hind Lamps Ltd., the concept was readily and contentedly accepted among the staff of Hind Lamps Ltd. All levels of staff participated in the inaugural function of Friends of Environment, as at that instance, their feeling was that, besides their routine job, they have a sense of responsibility towards environment and society. At the initial implementation phase of the activities of FEN, like environmental awareness rallies, organizing quizzes, painting competitions, plantation activities inside the campus, Hind Lamps Ltd. staff actively participated. Further when the implementation was phased out of Hind Lamps Ltd. and they have to move out in nearby rural areas, the feeling of responsibility was changed into burden and the staff started blaming to the initiators and promoters of FEN as they had

Figure 8.3 Stages of Group Development and Human Responses to Change

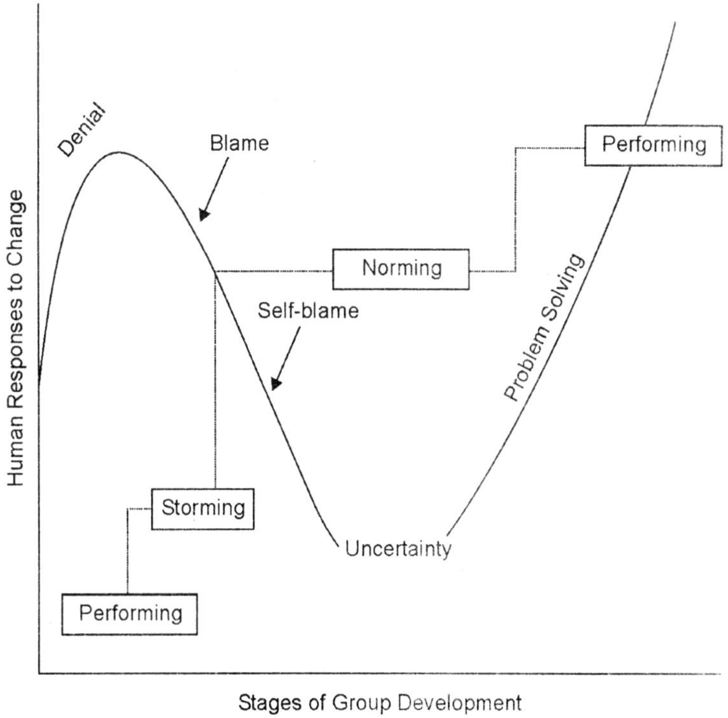

quite negative attitudes for local people who were presumed to be causing management problems for their organization. Employees developed a feeling that there is enough greenery in the campus so more plantation activities are not that essential. Repeated green initiatives may appear to be quite appealing in the beginning but at later stage it might not remain the same. They also felt that it is the responsibility of FEN staff to perform these activities. However, the involvement of the key functionaries, sometimes, forced the managers and staff to get involved. Providing new entrants with an orientation on the accepted values and culture within the organization and forming strategies, policies and procedures that are in line with the stated value system also ensures that those values percolated down to the lowest level of the organization (Purang and Sharma, 2007).

ENVIRONMENTAL VALUES

Overall, the managerial values for green initiatives create a picture that Hind Lamps Ltd. is contributing towards environmental protection and social development in a positive way. The organization believed that its first responsibility is to serve the society by delivering them environmentally safe products. It was observed that the top management of Hind Lamps Ltd. has shown positive attitude for development of environment-friendly products and systems as a long term strategic planning to build a better image in the market place. This study also indicates a positive change in the managerial behaviour of Hind Lamps as they felt a need to upgrade their machinery and products to fulfil the aspiration of potential consumers in future. The above discussion indicates that Hind Lamps Ltd. is an organization working for the social and environmental cause. In this regard leaders play a very important role for inculcation of right values. Peterson and Jun (2007) explored perceptions of small business managers regarding social responsibility in relation to various demographic and psychographic dimensions. A direct relation between degree of religious commitment and dedication to CSR concept was also found. The religious culture of organization was found to be instrumental in promoting well-being of society through self-sacrifice.

LEADERSHIP AND MOTIVATIONAL ISSUES

As far the analysis of leadership style, results revealed various functions and dimensions of leadership in Hind Lamps. Major findings indicated strong trend of participatory decision making as reported by the respondents. However, few policy level decisions used to be taken by consulting a few key functionaries. It was also observed that problems were considered as an opportunity to learn rather than hindering further work efficiency. Leaders played a major role in this regard by their, personal involvement as a leader while facilitating the problem solving process and patiently analyzing and managing the mistakes of their subordinates.

A need for more positive motivational environment at the organizational level was also felt as recognition and rewards for high performing personnel did not exist, which was against the motivational principles. Due incentives and reinforcement always increase productivity and satisfaction which is in the interest of organization, as positively motivated work force builds a healthy and reinforcing organization. Although leaders always believe in empowerment and capacity building of their staff, but such conscious efforts at the ground level are yet to take place.

The organization was found to be charged with positive as well as negative emotions. There was a need for more emotionally intelligent behaviour on the part of senior managerial staff in regard to addressing the challenging situations of local caste-based culture and political environment. Emotional intelligence is nowadays, prerequisite for playing an effective and efficient leadership role (Goleman, 1996).

To prepare a strategy, managers must decide where they want to be on the spectrum, from strict compliance to environmental leadership to rise to new heights of business of ethics along with production objective. In this regard leaders play a very important role in inculcating of right values. Current environmental issues span national, cultural, gender bias, educational, social and religious boundaries. Effective leadership requires knowing how to negotiate the issues that will enable to achieve vision and desired outcomes. Many people have built organizations that wither when the founding leaders leave after accomplishing the mission. But leaders expect to build a system, a system of values, culture and philosophy. This system was created by the veteran leaders of Bajaj Group of India, which percolated in generations of leadership to follow. The ethics and philanthropy in their soul is still encouraging the Bajaj corporate to transform their organizational units with a social and environmental mission. Transformational leaders operate from a personal value system that transcends their agendas and loyalties. In view of that, a more positive approach in practising transformational leadership are different levels may be required which can elevate the maturity level of the followers and convince them to strive for higher levels of ethical standards for social and environmental activities of their organization.

INNOVATING BUSINESS: ENVIRONMENTAL OPTIONS

However, the overall implications of environmental and social initiatives taken by Hind Lamps Ltd. could not have been convert into a great business opportunity had there been no vision, strategic direction and example emanating from the top executives of the company. Furthermore, coherence and integration will only be achieved, if the strategic plan indicates why CSR is important, what are its priorities, and the time scale involved. Besides, the appropriate level of resources allocation and how to achieve an appropriate balance between the present business, its potential and coping with the change for future, is also very crucial. The trading conditions and terms under which contractors and suppliers operate are likely to undergo considerable change. Companies like Hind Lamps Ltd., which already have green initiative programme in place, will be familiar with introduction of new quality constraints into the agreements with third parties. Additional constraints relating to sustainability could be the best way to achieve desired standards.

As stated by Porter (1996), how an industry responds to environmental problems may, in fact, be a leading indicator of overall competitiveness. Environmental regulation does not lead inevitably to innovation and competitiveness or to higher productivity for companies. Hind Lamps Ltd., has also been trying to innovate in terms of their green initiatives and has finally been successful to win the market place to some extent. This may lead Hind Lamps Ltd. to innovate further in lighting industry which is highly competitive. It can further enforce other players to take new standards as a challenge and respond to it with innovation. International competition has changed dramatically over the last two decades. Senior managers, who grew up at the time when environmental regulation was synonymous with litigation, are now considering that environmental improvement should be taken as good business opportunities. In a given situation Hind Lamps Ltd. would be better off by focusing on trade-off between cost and benefit instead of looking for win-win situation for all the stakeholders. In this case, environmental benefits should be weighed judiciously against tangible and intangible values of society and environment, as this will provide managers with information to set priorities and develop appropriate business response.

Some environmental issues are strategic because their impact on value is high enough either to put poor elements of the business at risk or to fundamentally alter a company's cost structure, and because managers have considerable discretion about how to respond. A good example is shifting to water-based coating in tube-lights produced at Hind Lamps Limited as against chemical coating to save the cost as well as to reduce the exposure to chemical which may be environmentally unsafe. Opinion is sharply divided on when and whether government regulation will prohibit the use of old chemical coating process. The value implication for Hind Lamp Ltd. are enormous not only because of the high cost of chemicals used in earlier technology to be saved, but also by virtue of their configuration to enjoy a relatively competitive advantage in this form of manufacturing.

As the situation suggests, one key decision that Hind Lamps managers must take about each major environmental problem they face is whether to lead or lag behind their competitors on environmental issues. In some cases, they may be interested to pursue an environmental strategy in which it gets well in the face of regulations or public opinion, and in other cases they may be served by moving with faltering steps with industry or by reacting only in response to external pressure. If Hind Lamps lag, they may find themselves on the receiving end of unfavourable regulations, but if they take initiatives and lead, their action may increase near their production cost in some respects and leave the company vulnerable to its competitors. Managers at Hind Lamps may find their options to be divided into those that help them shape events like forming partnership with stakeholders, and those that help them develop an optimal response to events, like re-allocation of resources and redesigning production processes. To prepare a strategy, the managers must decide where they want to be on the spectrum from strict compliance to environmental leadership.

CONCLUSIONS AND WAYS AHEAD

Looking to the projected high GDP growth of Indian economy, where industries are contributing substantially, there is a strong likelihood that it will further degrade the physical environment and put the people at more disadvantage in terms of environmental and societal development. However, the case of Hind Lamps Ltd. of Bajaj Group has very successfully practiced their CSR initiatives in business and thus demonstrated the vision of Group namely that business being a social institution has to try to fulfil its wider obligation towards the community in which it operates which is also in its own long-term and enlightened self-interest.

Another challenge that Hind Lamps Ltd. may face is in regard to their organizational system. Although they have the values for promoting participatory and healthy work environment, they may like to focus a little more attention on improving the motivational environment within the organization, encourage more transparency in communication among different levels and have more functional approach to conflict management. In this regard HR issues are very crucial to be addressed along with a policy for training and capacity building. It was realised that top leadership at organizational level is taking strong initiatives for optimal utilization of capacity and resources, and humanized approach to management through trust building mechanism, but how far such efforts will overcome the organizational inertia in terms of caste politics and attitudes of existing work force is an issue of concern.

Bajaj Corporate may also like to strengthen their flagship foundation for environmental initiatives, i.e. Friend of Environment in terms of its size, infrastructure and technical expertise if they want to broaden its scope and activities in future. This environmental expertise can be more optimally utilized by involving the staff of FEN in tackling the corporate environmental issues too as they have a challenging task of meeting the environmental requirements as per ISO 14001 which they have recently been awarded.

A green future is now being created by the visionaries of Bajaj Hind Lamps with competitive spirit in their belief and social actions. The products they wish to sell are not just consumables sold at a profit but are reflections of their corporate commitments to environmental concern and social responsibility. Leaders of Bajaj Group are sensitive enough to needs and problems of their employees within their constraints but need to provide more guidance and support to them is raising their level in transformational leadership continuum. This would help raise the motivational level of the workforce of Bajaj Group for societal commitments and organizational growth and development.

Above all, this study has created awareness among stakeholders of Hind Lamps Ltd. of Bajaj Group, regarding complex issues of business so that they can structure their views about strategies and techniques which they need to develop, in order to conduct the business in a way, which is consistent with environmental conservation and sustainable development. The consistent message by this effort is to continue to move ahead in the positive direction, and that is the only real commitment to change. It will further enable the Bajaj corporate to act effectively for business, society and environment in an ethical way.

It was realized that the impact of environmental and social initiatives needs to be much beyond just appreciation. However, whether and how strongly, they are going to integrate such philantrophic and environmental initiatives in their corporate strategy and how they are going to link it with market and equity, is an issue to look forward to.

Bajaj Electricals as a corporate has undertaken many social and environmental activities under their corporate social responsibility. Hence it is recommended that beyond reporting their financial gains they should also go for the GRI reporting pattern for their social and environmental initiatives to bring them on the national as well as international CSR map in terms of comparative assessment of these initiatives as per triple bottom line approach.

The consistent message by this effort is to continue to move ahead in the positive direction and that is the only real commitment to change. It will further enable the Bajaj corporate to act effectively for business, society and environment as per the concept of 'Be ethical'. The choice of Bajaj Electricals to embrace the planet earth and conserve its environment rather than consuming its resources unsustainably for corporate advantage and profit, is undoubtedly a noble idea which, in itself, is worth practicing, and will probably be an ideal for other corporate bodies to follow for equitable and sustainable business, society and environment.

NOTE

1. mrfcorporatesocialresponsibility.htm

REFERENCES

Arora, Garg and Vaida (2007) "Efficacy of Integrating CSR and Procurement Strategy: A Case Study of Leading Cotton Textile Manufacturing Company in India", *South Asian Journal of Management*, Vol. 14 (1), 105–119.

Bansal, P. (2002) "The Corporate Challenges of Sustainable Development", *Academy of Management Review* 16(2), 122–131.

Barney, J. (1986) "Organisational Culture", *Academy of Management Review*, 11(3), 656–665.

Carroll, A.B. (1991) "The Pyramid of Corporate Social Responsibility: Toward the Moral Management of Organisational Stakeholders", *Business Horizons*, 34, 39–48.

Charles and Thuard (2005) "Social Responsibility, Corporate Style". http://www.expresscomputeronline.com/20050502/technologylife 01.shtml

Crossley, D. (1999) "Paternalism and Corporate Social Responsibility Practices", *Journal of Business Ethics*, 21 (4), 291–302.

Daniel Goleman (1997) *Emotional Intelligence*, Bantam Press.

Ehrenfeld, J. (2005) "The Roots of Sustainability", *Sloan Management Review*, 23–25.

Elkington, J. (1997) *With Forks: The Triple Bottom Line of 21st Century Business*, London, John Wiley and Sons.

Emery and Oertel, (2006) "An Examination of Employee Culture-based Perceptions as a Predictor of Motivation": Proceedings of the Academy of Organisational Culture, Communications and Conflict, 11(1), New Orleans.

GEMI (2004) "Environmental Value to the Investors, Building Shareholder Values, The Global Environmental Management Initiatives", www.gemi.org

Gordon and Berry (1961) *Environmental Leadership: Who and Why*, Book BBR.

Gorte (2005) "Corporate Social Responsibility, Close to Victory", *Journal of Investing*, 14 (3), 140.

Hockerts, K. (1999) *Greener Management International* 25, 29–49, Spring.

Hubbard G. (2006) "Sustainable Organisation Performance: Towards a Practical Measurement System", *Monash Business Review,* Volume 2 Issue 3, November, 2006, www.sustainable.org-performance.pdf

Hunt, Hoskins, Schriesheim and Stewart. (1984) *International Perspectives on Managerial Behaviour and Leadership*, New York, Pergamon pp. 28–44.

Lantos. (2001) cited in Galliaria, D.M.A. (2004) *Derives and Models in CSR. Study Material in Corporate Social Responsibility*. Mumbai, Marsee Monjee Institute of Management Studies, pp. 47–59.

Mishra and Gupta (2007) "Using Semiotics to Understand the Culture of a Software Organisation", *Indian Journal of Industrial Relations,* Vol. 42 (3).

Mittal, M. (2007) "Corporate Social Responsibility: A New Way of Doing Business", *Management and Labour Studies*, Vol. 32 (2).

Peterson, T. and Jun, M. (2007) "Perceptions on Social Responsibility: The Entrepreneurial Vision", *Business and Society* online first, sage publication, http://bas.sagepub.com.

Porter, Michael E./van der Linde, Claas: "Green and Competitive: Ending the Stalemate". In: *Harvard Business Review*, September/October 1995, S. 12–134.

Purang and Sharma (2007) "Value Institutionalisation Enhances Commitment: Perspective of Culture in Indian Organisation", *Indian Journal of Industrial Relations*, Vol. 43 (1).

Schermerhorn, Osborn, and Hunt (2004) *Organisation Behaviour*, 8th edition, By John R Schermerhorn, Richard Osborn, James G. Hunt Reprint. Originally published by Wiley, C., 2003.

Smith, Wanda J., Richard E. Wokutch, K. Vernard Harrington, and Bryan S. Dennis (2001) "An Examination of the Influence of Diversity and Stakeholder Role on Corporate Social Orientation", *Business and Society*, 40 (3), 266–94.

Swanson, Diane L. (1995) "Addressing a Theoretical Problem by Reorienting the Corporate Social Performance Model", *Academy of Management Review*, 20 (1), 43–64.

Trainer, T. (2005) "Social Responsibility: The Most Important and Neglected Problem of All"? *International Journal of Social Economics*, 32 (8), 682–703.

Tuchman, B. (1965) "Development Sequence in Small Groups", *Psychological Bulletin,* 384–399.

Waldman, Siegel and Javidan (2004) "Transformational Leadership and CSR Working Paper"–0415, http://www.ideas.repec.org/p/rip/rpiwpe/0415.htm

Websites

www.babajelectricals.com

www.bsr.org

www.paryvaranmitra.com

www.thefinancialexpress.htm/February 05, 2006 / 'CSR is needed for our survival'.htm

Part IV

Environmental Leadership

EDITOR'S INTRODUCTION

Leadership is something unique, whether it is in environment or forestry or corporate or government or public sector. The common perception of leadership is governed by the above statement. However, the nature and style of leadership varies quite distinctively from one sector of organizations to another. A hardcore production organization may have different leadership centric requirements as against the softer sectors where service to society and community interactions are the integral part of organizational function.

In view of the above, environment is a sector where distinctive features of leadership are desirable for efficiency and effectiveness. This is the reason why Smithsonian Institution of Washington DC, USA, is running the environmental leadership course for international participants since more than a decade. The theme which is popularized is "create a world to which people want to belong" (Gilles Pajou). It is based on Robert Dilt's approach of visionary leadership and the Disney approach.

Certain salient features which make environmental leaders distinct from others may be the ethical consensus about nature and environment. Environmental leader always keeps his/her personal environmental values and ethics above everything. It gives him/her an upper hand in professional and societal requirements by serving the mother earth. The following two chapters primarily highlight such skills of environmental leaders.

9

The Challenge in Environmental Policy and Leadership

A Case of Bolivia

JUAN PABLO

The application of environmental policies is a key factor for biodiversity conservation. The ecological problems are not only social and economic. The political stability of the country and the adequate implementation and application of legal regulations drives sustainable development. Basically, the processes are more important than a site threat, and the environmental policies are processes that need to be considered seriously to achieve adequately conservation goals. The questions still are: Who is going to lead those processes? How can we keep a strong leadership in this changing governmental organization and decision-making process? How could society embrace the complex concept of sustainable development, if for each new administration we get new Environmental Ministry structure, political instability to follow up environmental and conservation trends, and inadequate indicators that could help the analysis of sustainable development measures?.

This chapter primarily assesses the weaknesses of the environmental leadership towards a consistent agenda, towards a sustainable environmental policy in developing countries. As a case study, the author explores the historical background of Bolivia, characterized by a political instability in the governmental structure and organization to achieve environmental goals from 1989 until 2007. Despite the historic context and environmental achievements in the protected areas system, the sustainable development framework, local participation, and global environmental challenges, it is expected that new governmental administrations in Bolivia could address new environmental policy and leadership opportunities.

Several studies on public policies have been carried out considering basically social and political issues in Bolivia. According to David Gambill (1999), the knowledge of property rights by community strength the sustainable use of natural resources provides a good chance to enforce the application of environmental policies. In the same way, the corruption can be decreased and the community is open to enforce the rules and regulations to implement best practices in a sustainable development manner.

The controversial issue between environment and development is still a problem in Bolivia, and many other countries. According to Kaimowitz et al. (1998), the decentralization process in Bolivia is a good opportunity for municipal governments to enforce regulations, leadership and to achieve benefits from forest resources and policies. The General Environmental Law as a legal tool to achieve sustainable development is still weak because the relative high pressure for high economic growth (Blacutt-Mercado, 1993).

The economy could have long-term effects on the environment generating employment and income distribution. But, detrimental failures against the environment is due to inadequate incentives for sustainable development causing market problems, expansion of exporting natural resources, trade liberalization and others. The environmental relationship with economy and vice versa is part of the whole sustainable development policy to address development and social welfare.

For example, the implementation of environmental policy in Bolivia and other developing countries as well, depends on many factors, due its notable cultural, linguistic and environmental diversity and social and economic differences within the country. The municipal level of decision making and the protected areas are key issues to enhance the implementation of environmental policy in a context of a sustainable development framework. Since the creation of the Ministry of Sustainable Development and Environment in 1993, the environmental guidelines have been changed from one administration to another. Thus, most of the environmental, biodiversity conservation, and natural resources management norms have been controversial at municipal level. A strong leadership and guidance can increase the understanding of nature as the best investment that motivates an effective implementation of environmental policies.

At present, Bolivia is challenged to consolidate its long term sustainable development policy. The dramatic global economic recession it is affecting Bolivia's economy as well as its environmental policy and sustainable development strategies. In the rural areas, the social pressure on natural resources is resulting in a vicious circle of "greater poverty causes greater ecological problems and deterioration of nature and the quality of incomes get worse and again cause greater poverty". The change of political administration, environmental priorities and conservation goals are permanent political constraints that affect the environmental behaviour and attitudes of the people and institutional relationship among government, private sector and civil society.

Thus, local populations are exploiting natural resources in unsustainable manner, due to inadequate leadership for the implementation of environmental policy and sustainable development initiatives. It is important to develop alternative solutions to get investments and social opportunities in order to achieve a political stability of the local government, and moreover, a healthy environment, adequate developmental priorities, and a sustainable human development as a whole.

THE GOVERNMENT STRUCTURE SYSTEM, ENVIRONMENTAL POLICY AND LEADERSHIP

Bolivia like many other countries in Latin America, has been implementing political reforms since the military regimes collapsed at the end of 1970s. The transition to democracy between 1978 and 1982 was violent and with economic consequences to the people. The first democratic coalition in 1986 under Siles Suazo administration was characterized by a struggle of economic reforms and political instability. Only a few years later the neo-liberalism took over the governmental modernization of the state given market liberalization and price control abolition, open foreign trade and capital investment and liberalization of labour market (Mayorga, 1997). The political influence of IMF and World Bank on governmental decision-makers was remarkable during the 1980s.

The concept of sustainable development became important in the 1990s. It involves a vision of security, equity and sustainable future for the human kind. The United Nations Conference on Environment and Development (UNCED) in Rio de Janeiro defined it as "economic growth that meets human needs while protecting nature's ability to renew itself". The International Forum on National Sustainable Development Strategies (NSDS, 2001) held in Ghana in 2001 debated lessons learned, country experiences, and assessment processes to implement sustainable development programmes. Based on these principles, sustainable development includes environmental policies in the framework for capacity building and participation. In Bolivia, the Popular Participation Law and other legal initiatives assess this important strategy, but with a very weak political support, lack of knowledge of rural level problems, economic constrains, and incipient democracy.

Since democracy returned to Bolivia in 1986, the governmental agenda was to mitigate the critical economic crisis of the country. For the Minister, the highest political decision maker, sustainable development was a learning processes in itself, because in the past, most of the developmental economic plans in Bolivia were mainly based on economic growth, security, and production. Especially during the dictatorship period (1970–1985), environmental and sustainable use of natural resources, were weak strategies of economic developmental plans.

In Bolivia the environmental efforts began in 1939, where President German Bush promulgated a Supreme Decree, to regulate forestry concessions, mandatory re-planting activities and a mechanism to establish wildlife protected areas. During the first administration of President Hugo Banzer in 1975, another legal norm was promulgated: The Wildlife, Hunting and Fishing

Law (Ley de Vida Silvestre Caza y Pesca). This law was important to strengthen the wildlife protection and to preserve watershed catchments. The governmental structure to support this initiative was the Natural Resources Division, under the Forestry Service and finally under the Ministry of Agriculture. By the time, in the early democratic years of Bolivia in 1989 during the administration of Jaime Paz Zamora (1989–1993), new initiatives came out. Taking into consideration the unsustainable use of natural resources causing a potential risk for biodiversity and ecosystems in the country, in 1990 the government determined the 'PAUSA ECOLOGICA HISTORICA', an environmental policy regarding a complete suspension of forestry concessions, fishing, and hunting activities. Since 1952, this policy was a reacting attitude against the traditional exploitation of natural resources and economic development. The Paz Zamora administration, engaged the nation to start a new development model as a transition process and governmental transformation of the already new democratic life in Bolivia. During this government created the National Trust Fund for the Environment (FONAMA) and the General Secretariat for the Environment (SENMA) with higher power of political decision within the executive structure. This higher status of the public environmental offices may be the most important initiative in this matter. The international agencies and cooperation in Bolivia described this new structure as a good opportunity to allocate funds for programmes and projects. The new General Law of Environment was promulgated in 1992 giving the chance to all organizations to be part of the new environmental and sustainable development process, with a complete representation of all sectors, such as NGOs, private, academic and communities at national, departmental and local level.

Since the creation of the Sustainable Development and Environment Ministry in 1992, what effects of sustainable development learning are occurring in Bolivia? To answer this question, it is necessary to go back to two previous governmental administrations and understand that the learning process in sustainable development issues in Bolivia is very close to the creation of the Ministry of Sustainable Development. A big change in the Bolivian political life was the establishment of the National

Plan for Social and Economic Development (PGDES) in 1992, based on a sustainable development framework. Currently, the new government administration is working on this concept again, but with a new and powerful institutional organization and authority of the ministry.

After the UN Conference on Sustainable Development in Rio in 1992, a complete change of political perspective in Bolivia's young democracy took place. Thus, the President, Sanchez de Lozada, created the Ministry of Sustainable Development and Environment (MDSMA) as the main governmental strategy and institution empowered to link economic development and human development objectives.

This transmission of sustainable development experience and knowledge from one administration to the next was an important approach in the sustainable development learning process in Bolivia. It makes clear that the learning process for the past nine years was maintained by the technical structure of the ministry. There were officials interested in learning more about sustainable development, and maintaining strategic linkages with bilateral and multilateral middle officials as well.

International financial organizations, such as bilateral and multilateral donors, for instance, were comfortable in having a national guideline based on sustainable development, better suited for their economic support. In the Bolivian experience after RIO 92, the learning process about sustainable development increased in the society as a whole. The Global Environmental Facility (GEF) from World Bank is an example of this learning process. Once the National Parks System (SNAP) was established, the World Bank was interested in enhancing a long-term sustainable financial programme. The Netherlands Protected Areas Programme for Bolivia (PASNAPH) donated $ 20 million for the next 20 years to SNAP, contributing to consolidation of the National Park Service and SNAP. This learning process of sustainable development favourably improved an endowment strategy for supporting the national parks system. How? The middle technician's relationships with donor officials enhanced and clarified the sustainable development framework for critical economic support. In fact, the donors understood, not only the learning process in sustainable development, they understood also the practical issues behind this learning process: to allocate funds in a positive and effective way.

During the first years of administration of Sanchez de Lozada (1992–1997), the organizational structure of the MDSMA, was a new initiative for donors, civil society, among other agents of learning in our policy network (Figure 9.1). He established five major vice-ministries (see Figure 9.1); the most important was the Vice-Ministry of Popular Participation. In sustainable development learning, popular participation is the main strategic tool for achieving knowledge and sharing experiences in sustainable development. After the decentralization of municipalities in 1993, the concept of sustainable development became a critical strategy for local communities. Under a new social reality, politically well supported, and economically autonomous, the municipalities have been learning the concepts and benefits of sustainable development.

The Vice-Ministry of Indigenous Affairs and the Vice-Ministry of Natural Resources and Environment, have developed new communication systems within the civil society in order to get acceptance, consensus, and feedback to continue the learning process. For instance, the office of Sustainable Development and Promotion was invested with the power of disseminating the concept of, and the economic and social implications of sustainable development. It affected bilateral and multilateral donors, civil society, private sector, and local interest groups. Within the ministry, this office was empowered with organizing the learning and training process on sustainable development to

Figure 9.1 Establishment of Sustainable Development and Environment Ministry (MDSMA: 1992–1997)

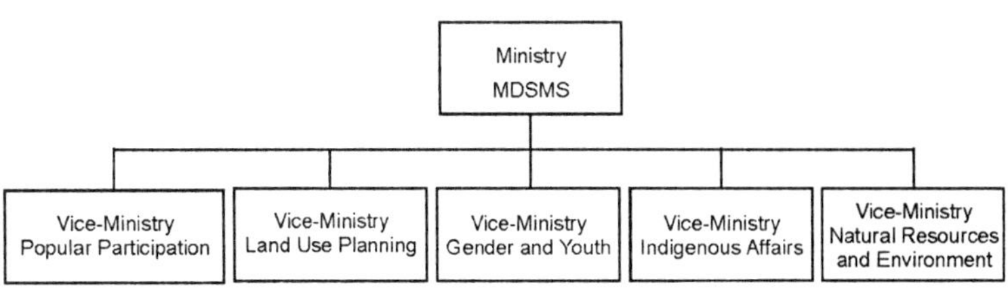

technicians, policy middle men, and lower level decision makers. Unfortunately, during the administration of Banzer Suarez (1997–2002), this process did not have any adequate continuity and political support.

Basically, during the Banzer Suarez administration (1997–2002), the sustainable development concept did not have a place on the governmental agenda. Natural gas, poverty alleviation, and narcotics traffic were the main political guidelines. The name of the ministry was changed to Sustainable Development and Planning (MDSP) with three vice-ministers in charge of Land Use Planning/Popular Participation, Gender/Youth, and Forest Exploitation and Natural Resources Management (Figure 9.2). A practical political decision and adjustment, resulting from the learning process in the organizational structure and role of the ministry, was the decentralization and creation of the National Park Service (SERNAP); an autonomous institution in charge of the sustainable development of the national parks system. Another example is the relationship between middle technicians with the Intergovernmental Panel on Climate Change (IPCC). With the financial support of The Netherlands, GEF, United Nation Development Programme (UNDP), and the US Country Studies Programme, these contacts enhanced the consolidation of the climate change plan in Bolivia.

The Ministry of Sustainable Development and Planning during the current second term of Sanchez de Lozada

Figure 9.2 Sustainable Development and Planning Ministry (MDSP: 1997–2002)

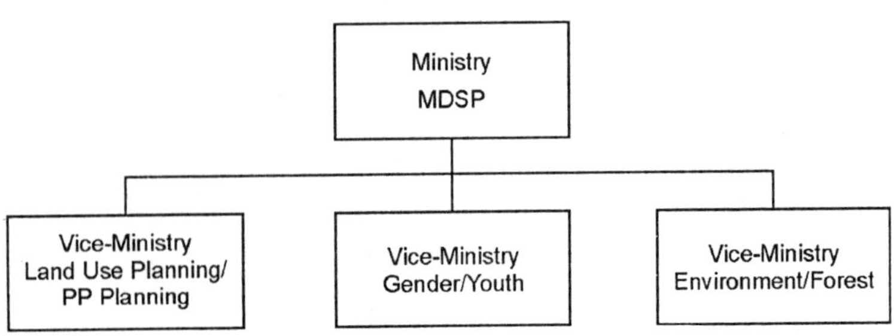

(2002–2003), from the MNR political party, had a new organizational structure (Figure 9.3). There are three Vice-Ministers: Political Economy, Land Use Planning, and Natural Resources/Environment. Our analysis reveals that the National Parks Service and the National Climate Change Programme, both still autonomous, are the main offices where sustainable development has been fundamentally learned. Thus, both will have an important role in the sustainable development learning process within the new structure of the ministry.

The Ministry of Sustainable Development and Planning now has greater authority and political power. This is an opportunity and a challenge for moving forward sustainable developmental strategies in the country. According to Hall's learning orders, I consider that fundamental changes in goals and objectives are coming (third order learning). Moreover, the conceptual role of the new Vice-Ministry of Political Economy will link the economic power of the MDSP (budget issues) with sustainable development and social welfare in a better way. The minister is also head of the economic council, a key group of ministers who make final decisions on governmental agenda.

In the Richard Rose's work on lessons-drawing and avoiding policy problems, decision-makers can have three alternatives: "to turn to their national past; to speculate about the future; or to seek lessons form current experience in other places" (Bennet and Howlett, 1991). The current political coalition, between MNR and

Figure 9.3 Sustainable Development and Planning Ministry (MDSP: 2002–2003)

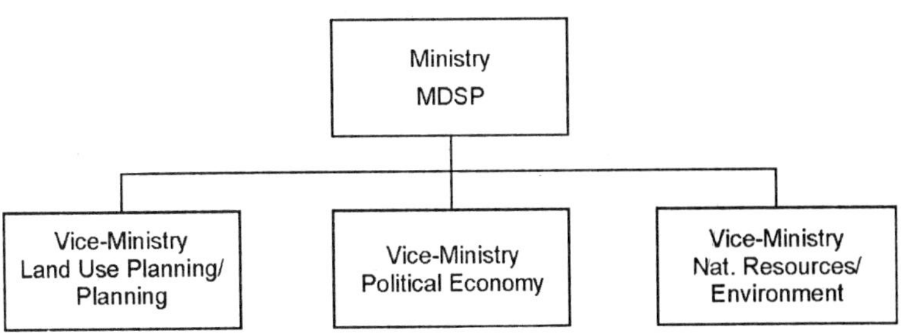

MIR, has a crucial sustainable development learning strategy from the past.

A good example of this challenging and controversial situation is described using a simple gamma correlation coefficient (A gamma correlation coefficient measures the association between variables) in the distribution of the government system variables and the environmental norms. In Table 9.1 and 9.2, the environmental achievements and protected areas support by each government administration is described, as well as their correspondent ordinal ranks.

Table 9.1 Environmental Achievements and Protected Areas Support

	Environmental achievements			Protected areas support		
Government	**High**	**Medium**	**Low**	**High**	**Medium**	**Low**
1989–1993	1	10	7	0	3	3
1993–1997	6	7	5	2	4	1
1997–2002	4	6	8	2	4	1

Source: Arce, 2002.

Table 9.2 Ordinal Rank for Environmental Achievements and Protected Areas Support

Environmental achievements	Protected areas support			
	High	**Medium**	**Low**	**Total**
High	2	3	3	8
Medium	2	3	3	8
Low	2	3	3	8
Total	6	9	9	24

Source: Arce, 2002.

The gamma correlation coefficient is Ns= 78 and Nd = 63. The resulting gamma coefficient equals to 0.1. It shows an interesting correlation between the governmental system of variables. This correlation shows the governmental instability regarding environmental achievements and protected areas support, funding, and management. During the Paz Zamora, Sanchez de Lozada and Banzer administrations, the environmental policy had different approach approaches. For instance, in 1989–1993 environmental mitigation against mineral pollution and biodiversity protection were the main issues. During the Sanchez de Lozada administration (1993–1997), sustainable development, protected areas, environmental impact evaluation, and climate global change, among others, were environmental priorities. During the Banzer/Quiroga administration (1997–2002), some priorities were put on protected areas, climate global change, energy resources and starting special norms to land tenure rights. There are no a clear environmental policy consistency and leadership along all administrations. Although, Bolivia addressed Agenda 21 ten years ago, sustainable development paradigm is still an ambiguous concept in the Bolivian society. The learning process as mentioned before is socially and environmentally, the major challenge and achievement by civil society and government.

Unfortunately, Bolivia had a dramatic political climate, and the elected president Gonzalo Sánchez de Lozada resigned in October 2003. He was substituted by a short period of time by Vice-President Carlos Mesa. Mesa, and in turn replaced by Chief Justice of the Supreme Court Eduardo Rodríguez in June 2005. In less than two years the political leadership of the country had the major challenging situation for its democracy.

The 2005 Bolivian elections, the Movement Towards Socialism (MAS) won the election in an absolute majority unusual in Bolivian elections. The new President, Evo Morales was sworn in January 2006. Although the first year of the new administration focussed on re-activation of the economy, employment, and the natural gas strategy; the consolidation of land tenure, natural resources management, and biodiversity conservation. These policies are crucial considerations in avoiding future social, economic and environmental problems. This is the case especially with the promulgation of the nationalize Bolivian hydrocarbon assets, and the Assembly began drafting a new constitution. There is an interesting chance and a source of hope to achieve environmental policies more in depth with the conformation of new

Figure 9.4 Ministry of Agriculture, Rural Development and Environment (2006–2008)

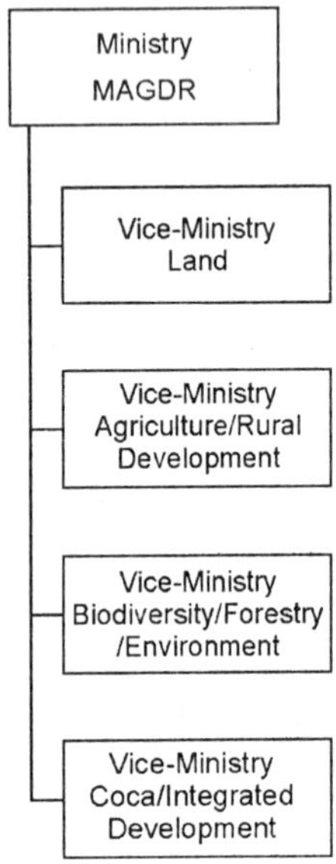

Figure 9.5 Ministry of Planning and Development (2006–2008)

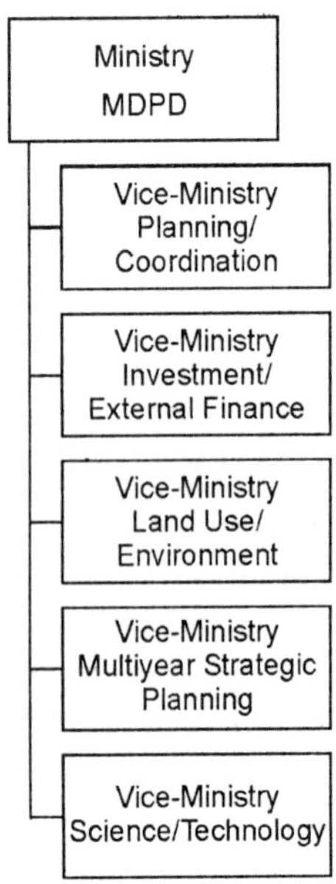

political coalition in the benefit of a unique environmental agenda.

However, the "Sustainable Development and Environment" concept changed again and switched onto two new Ministries according to the new governmental structure (Law 3351, February 2006). The new ministries are, the Ministry of Agriculture, Rural Development and Environment (MAGDR), and the Ministry of Planning and Development (MDPD). It is important to mention both, because there is a complex relationship in the roles and leadership on the environmental and sustainable development policy issues. In the Figures 9.5 and 9.6 both organization structures are described—showing up the new decision making process on its role.

Taking into consideration all of these considerations and challenges, I am optimistic that in Bolivia or similar situations in other developing countries, the decision-makers can develop a strong leadership in order to address sound environmental policies. Some considerations for the future can make significant difference as discussed below:

SOME CONSIDERATIONS FOR THE FUTURE

Empowering Leadership at Local Level

The role of municipality in protected area management is still a challenge. The integration of the municipal development plan into the protected area management plan is crucial. Who comes first? Are municipalities within protected areas or protected areas are within municipalities? This is a key question in building a strong leadership at local level. Both institutions require political support and sharing.

The National Protected Areas Service

The institutional framework to administrate the system is a national authority, the National Protected Areas Service (SERNAP), established by Supreme Decree 25,158 in 1998. Actually, the SERNAP is no longer a decentralized institution. It belongs the Vice-Ministry of Biodiversity, Forestry and Environment. In fact, the protected areas system is a natural space that should be considered in the land use planning of the country, due to its importance in achieving sustainable development at national and regional levels of analysis.

The National Protected Area Service should be a decentralized and technical institution. In our experience, this is the better way to strengthen the local leadership, conservation investments and the implementation of adequate environmental policies.

The Sustainable Development Council

The future of Bolivia with its new natural gas economy, oil exploitation, and a vast land of high biodiversity values and natural resources, depends on how much the people will learn from current experiences in other places and themselves. It is time to restart that the Sustainable Development Council, a civil society participation mechanism created around 1993–94, can strengthen the sustainable development learning process and sharing initiatives as a whole. Its is crucial to show up a strong leadership and commitment to share ideas.

The Biodiversity Conservation and the Protected Areas Bill

These two legislative initiatives are still under consideration since 1993. It requires a political commitment and willingness for sustainable development and social welfare in Bolivia. These norms are the key tools to consolidate a strong leadership at governmental level.

A New Sustainable Development Paradigm?

The new political scenario in Bolivia under the MAS administration and the new political forces of the opposition represent a historical opportunity for the minorities, farmers and ethnic groups to change the sustainable development concept. It is a challenge for the working class to maintain a positive political attitude to build a strong environmental leadership. In this complex political culture, a source of hope is to avoid any power or elite temptation, for example the compromise of MAS to environmental and social issues, such as natural resources, sustainable development, sovereignty, employment and human rights policies (MAS, 2002).

CONCLUSION

This contribution assesses the weakness of the environmental leadership towards a consistent agenda towards a sustainable environmental policy in developing countries. As a case study, the author explores the historical background of Bolivia, characterized by a political instability in the governmental structure and organization to achieve environmental goals from 1989 until 2007. Despite the historic context and environmental achievements in the protected areas system, the sustainable development framework, local participation, and global environmental challenges, it is expected that new governmental administrations in Bolivia would address new environmental policy and leadership opportunities.

ANNEX I

Comprehensive Data by Municipality

Information	Sources	Place
Environmental Policies (1989–2002), environmental organization	Capacity 21 documents Ministry of Sustainable Development League for Environment (LIDEMA) Conservation Policy—FSD	BOL BOL BOL DC
Government Strategies for Social and Economic Development (1989–2002)	Ministry of Communication Official Gazette	BOL BOL
Laws, regulations, norms, structure and budget	Official Gazette National Statistic Institute Ministry of Presidency	BOL BOL BOL
Environmental programmes and zoning	Conservation International-Bolivia SERNAP	BOL BOL
Amboro-Madidi CI's Conservation Strategy	Andes Programme CEPF/CI Field Support Division FSD/CI Conservation International-Bolivia	DC DC DC BOL
National Parks: area and maps	Conservation International-Bolivia and National Parks Service (SERNAP)	BOL BOL
Indigenous territories (TCOs): area and maps	Conservation International-Bolivia and League for Environment (LIDEMA) National Confederation of Indigenous Communities (CIDOB)	BOL BOL BOL
Departments, Provinces and Municipalities maps	Conservation International-Bolivia and National Statistic Institute	BOL BOL
Land use change maps (1990–2002)	Conservation International-Bolivia	BOL
Roads map	Conservation International-Bolivia CABS/CI	BOL DC
Agricultural Areas and Colonization	Conservation International-Bolivia	BOL
Forestry Concessions	Conservation International-Bolivia	BOL
Population Census 2001 per province	Conservation International-Bolivia and National Statistic Institute	BOL BOL

Cont'd...

Cont'd...

Population Census 1992 and 2001 changes	Conservation International-Bolivia and National Statistic Institute	BOL BOL
Housing	National Statistic Institute	BOL
Poverty Index Census 1992 and 2001	National Statistic Institute	BOL
Education and Health services	National Statistic Institute	BOL
Province investments	National Statistic Institute	BOL

REFERENCES

Arce, J.P. (1990) Environmental Impact of Beef Cattle Activity and Growth Centres Analysis for Conservation Planning (Department of Beni, Bolivia). An Application of GIS and Environmental Impact Assessment Techniques. M.Sc. Degree Thesis, ITC, The Netherlands.

Asquit, N. (2002) "How to Prioritize Policies for Zero Loss". Presentation to CI Annual Meeting April 26th, Washington; *Conservation International.*

Bennet, C.J. and Howlett, M. "The Lessons of Learning: Reconciling Theories of Policy Learning and Policy Change". *Policy Sciences*, 25 (3) (1992), 275–294.

Blacutt-Mercado, W.P. (1993) "Environmental Legislation, Economic Growth and Risk in Minerals Development: The Bolivia Case". *Natural Resources Forum,* 17, 207, Abstract from *Infotract:* Article A14577993.

Bolivia.com. "Electiones Generales Bolivia 2002". La Paz/Recta Final. CNE. 7 de Junio de 2002.

Earth Council (2001) The Rio+5 Consultation March 13–19, 1997, Retrieved May 21, 2001, from the Worldwide Web http://www.ecouncil.ac.cr/rio

El Diario. "MNR y MIR Firmaron Acuerdo por Gobierno de Responsabilidad Nacional", La Paz/Noticias El Diario-Politica. 26 de Julio de 2002.

Fundacion Milenio, "Informe de Milenio sobre el Desarrollo Sostenible en Bolivia", *La Paz:* Fundation Milenio, Noviembre 2001.

Gambill, D. (1999) "Intentionally Sustainable: How Community-based Resource Management Enables and Encourages the Sustainable Use of Resources". Gender Working Paper 1, 1–8.

Kainmowitz, D., Vallejos, C., Pacheco, P.B. and Lopez R. (1998), "Municipal Governments and Forest Management in Lowland

Bolivia. *Journal of Environment and Development*, 7, 45, Retrieved June 1, 2002 from the World Wide Web: http://web6.infotrac.galegroup.com/itw/infomark/770/392/23370874w6/piurl=rcl_EAIM_0_A20798471&dyn

La Razon. "La Organization y los Planes de Gonzalo Sanchez de Lozada". *La Paz:* Grafico J.H.T./*Periodico La Razon*. 24 de Agosto de 2002.

La Razon (2002) "20 Anos de Democracia" Revista Zoom, Capitalization;
(a) Nuevo Modelo, "Corruption", *Pactos Los hijos de la democracia. Periodico;*
(b) *La Razon.* 18 de Octubre de 2002;
(c) http:/www.la-razon.com/Zoom/Octubre/zoom_oct_2002g.html

MAS (Movimiento al Socialismo) (2002) "Territorio, Soberanía, Vida. Programa de Gobierno", *Movimiento al Socialismo*, Bolivia.

Mayorga, A. (2001) "Bolivia: Electoral Reform in Latin America", IFES, UN-DESA, IDEA. From http.aceproject.org/main/English/es/esy_bo.htm

Mayorga, A. (1997) "Parliamentarized Presidentialism, Moderate Multiparty System and State Transformation: The Case of Bolivia". *Journal of Democracy* 7, p. 4.

Ministerio de Desarrollo Sostenible y Planification. (2000) Primera Comunication Nacional ante la Convencion de Cambio Climatico. Edicion N. Roca H. Viceministra de Medio Ambiente Y.O. Paz R. Coordinador Programa de Cambios Climaticos. Bolivia.

Ministerio de Desarrollo Sostenible y Planification. Informe de Gestion (2002) Redaccion y Edicion J. Oporto V. Y.A. Lino C. *Unidad de Comunication y Prensa*. Bolivia.

Molina, S.M. (1993) "Caminos Hacia la Sostenibilidad. Participation Popular, Herramienta del Desarrollo Sostenible". *Capacity 21 Estudio Nacional*, Bureau for Development Policy, New York.

NCSD Knowledge Network, "NCSD Rio+10 Assessment—Process Agenda (2001)", National Council on Sustainable Development. WSSD/Rio +10. Retrieved May 2, 2001 from the Worldwide Web http://www.ncsdnetwork.org/knowledge/events01.html

Servicio Nacional de Areas Protegidas. (2002) Areas Protegidas de Bolivia, Conservando la Biodiversidad: *Memoria* 1998–2002. Sernap-Ministerio de Desarrollo Sostenible y Planificacion/GTZ/Banco Mundial, Bolivia.

United Nations (2001) Johannesburg Summit 2002 from 26 August to 4 September, Retrieved May 2, 2001 from the World Wide Web.

10

Essential Skills for Environmental Leadership in a Global Context

LESLIE E. WILDESEN AND JEANNE C. CROUCH

Current environmental issues span national boundaries, cultures, income levels, genders, degrees of development, education levels, languages, and religious frameworks. Everyone on earth needs clear water, clean air, healthy food, sturdy and culturally appropriate shelter and clothing. Increasingly, people need culturally and environmentally appropriate transport, communications, and inclusive ways of organizing society, economy, education, and religion. Who will lead the earth's 6 billion people in their quest for a sustainable future? What tasks will they be called upon to perform? What skills will be necessary for these leaders? How can they learn and practice these skills?

This chapter is not intended as an academic treatise. Rather, we will explore these questions from a perspective gained from working with participants from around the world in The Smithsonian Environmental Leadership Course, which Wildesen and Crouch designed and developed, and which, with Dr Francsico Dallmeier, Director of the Smithsonian's Monitoring and Biodiversity Programme, has been delivered for the Smithsonian Institution (USA) since 1997.

WHO ARE THE LEADERS?

Those who will lead the way are a varied lot. They are scientists, politicians, journalists, teachers, consultants, heads of (and members of) NGOs, religious figures, and concerned citizens of all nations. They span all age groups and they all expect to lead vigorous lives for many decades to come. They have doctorates, masters' degrees, baccalaureates, or no formal education beyond secondary school. They like to hike, dive, sail, run, play volleyball and ping-pong. They play the guitar or the piano or the drums, they dance, sing, act, draw, paint, and take photographs. They are from Africa, and Asia, and Latin America, and Australia, and the Pacific, and Europe, and North America.

They are Hindu and Muslim and Jewish and Christian and Buddhist and non-religious. They are from republics and democracies and kingdoms and dictatorships and all the political variations you can think of. They are from rich families and poor families; from the cities and the countryside; from the tropics and the icy north; from continents, islands, and islands the size of continents. From nations with millennia of traditions, and from nations that gained 'nationhood' within the past 50 years.

Some are widely known, some are not. But they all are doing important work, sometimes with almost no staff or resources. They are courageous. They are creative. They are humble. And they are already being successful.

WHAT TASKS DO ENVIRONMENTAL LEADERS PERFORM?

They raise funds. They hold meetings. They negotiate contracts. They prepare applications. They write articles. They give speeches. They inspire, motivate and lead their staff, their colleagues, their followers, their public, their nations. They create whole programmes out of nothing but their imagination and determination: national forest systems, protected areas, eco-tours, native plant nurseries, 'save the...' campaigns, endangered species studies; programmes to fight poverty, promote health, educate about the environment, encourage participation, ensure justice; programmes funded by the United Nations, the World Bank, national or local governments and NGOs, grants, donations, and sweat equity. Then they get up the next morning and do it all again.

WHAT SKILLS ARE ESSENTIAL FOR THESE LEADERS?

We believe with Robert Dilts (1996) that leaders in whatever arena—organizational, political, and especially environmental—need the ability to function well in four different but related skill areas: (i) self-skills, (ii) relational skills, (iii) strategic skills, and (iv) systemic skills. These skills, added to whatever subject-matter or professional or topical skills they may have, such as biology, or law, or writing or business management, along with a large dose of creativity, enable the leader to bring their ideas to fruition in the real world.

Self-Skills

Self-skills are about how you conduct and manage your own behaviour in various situations. These skills enable you to *select* the attitude and emotional state that will be most effective in entering a situation and to generate that attitude within yourself. A keypoint about these self-skills is that you can *choose* the state of mind and attitudes that you will utilize. You do not ever have to let your attitudes and internal state be a random event (i.e., whatever state you were in when you woke up that morning). It is our experience that increasing one's self-knowledge and improving one's self-management skills can dramatically improve one's effectiveness as a leader.

You need to know yourself. Several ways to do this include: understanding the ways in which you have typically taken in and processed information (visually, auditorally, kinaesthetically); how you have approached various aspects of your personal environment (people, places, things); and how you typically approach a project or activity (do you move away from problems or towards a goal? look for small details or at the big picture? focus on the past or on the future?).

For example, many people who consider themselves to be leaders focus on the big picture of the future, and move towards this goal actively. This is the classic profile of a visionary, and these are wonderful personal traits, as long as the person understands that it is necessary to have an assistant (or many of them) who can handle the small details, make sure problems get solved, and ensure consistency over time in approach, tone, and quality of action. Leaders who do not recognize the need for this will flounder (usually sooner rather than later) and will never be able to sustain their 'big idea', no matter how worthwhile it is.

On the other hand, many environmental scientists tend to focus on the details, and design their studies to respond to past studies and findings, for the purpose of finding the best explanation for the current situation. This too is a wonderful set of traits, and it is extremely important for someone with these skills to team with people who can easily keep focused on the overall goal of the project, rather than getting caught up in the details. The scientists who understand their specific area of expertise but who fail to connect with 'big-picture' people who can link the specifics to a larger scale plan and adequate resources will also be unable to create the changes they seek. You also need to hone your perceptual skills and become aware of the impact of your behaviours and attitudes on others. Further, developing the flexibility to be able to modify your behaviour and style to the situation is crucial for leaders. It is not enough to simply understand yourself, although this is a crucial first step. You must also be able to step into another person's shoes, to understand and value their needs, motivations and ways of approaching the issues at hand which brings us to the next category, relational skills.

RELATIONAL SKILLS

This set of leadership skills is all about how you communicate with others, and the depth and effectiveness of the relationships you are able to build. Clearly, any leader must influence others to achieve his or her outcomes. Leaders persuade, motivate and inspire others. They catalyze followers/team members/employees to action. If you are to reach your vision, you must have highly developed relational skills.

And it is important to realise that those 'others' with whom you communicate perhaps—probably—will not share your same personal profile or values, beliefs and work style, nor will they be aware, oftentimes, of how such things affect their behaviour. Therefore, the 'leader' often must do all the work (or a large part of it) in developing and applying these understandings.

For example, if you are trying to gain support for your environmental project or programme it is your responsibility to prepare the briefing materials in a format that can be readily understood and appreciated by your audience. If the boss (or the prime minister or the UN delegate) wants to see numbers in a spreadsheet, do not go into the meeting with a narrative text and some slides (and vice versa). If you are anticipating a large meeting with a lot of different people and organizations, make sure you are prepared for all of the possibilities, and have materials that respond to many contingencies.

The leader must be skilled at developing rapport with others. There are very simple techniques to do this. Some leaders have a natural ability to do this, others need to enhance or develop their rapport skills. This can be done by learning the physiological, tonal, and linguistic basis of rapport.

Great leaders use these skills to create a relationship, then build that relationship through demonstrating their own competence and by deepening their understanding of the background, style and worldview of those they wish to influence. Again, some leaders are naturally skilled at stepping into the world of others. Others must first learn its importance and then learn how to do it. Once one has developed this skill it becomes much easier to develop or negotiate a strategy that meets the needs of both the leader and those s/he wants to lead or influence.

In fact, self-skills and relational skills taken together form the basis of what Daniel Goleman calls 'emotional intelligence' (Goleman 1997, 2000; Cherniss and Golman, 2001). In our experience working with leaders from all walks of life and in observing leaders on the world stage, it is the combination of these two sets of skills that separate leaders who are *both admirable and effective* from those who are *merely effective*. In other words, leaders who have strong self and relational skills are the leaders others want to emulate which brings us to strategic skills.

Strategic Skills

Effective leadership requires knowing how to chart the course or navigate the path that will enable you to achieve your vision and desired outcomes. Self and relational skills are not enough. You must be able to evaluate where you are with respect to where you want to be, and what will be required (which people and what actions) and when in order to move closer to your desired outcome.

Planning, designing, negotiating, scheduling, prioritizing, setting goals and outcomes—all these tasks require strategic skills and tools. The higher up in your organization or programme or public visibility you go, the more you need to be able to develop goals that are achievable, budgets that are realistic, programmes that work for all the players, and measurements that everyone can understand and apply. You need to develop strategies for discussing controversial topics, and for generating support among opponents—and for doing the next steps if you cannot generate the needed support at this time. You need to figure out how to stay motivated, and how to keep your staff, colleagues, and followers motivated, especially when the situation may not be going your way. And you need to create a strong system for whoever comes after you to keep the project or programme going which brings us to systemic skills.

Systemic Skills

Systemic thinking is the ability to understand the system (not just the specific situation) in which you are operating. This includes the people, culture, legal framework and rules, budgets, the physical environment, the economy, and the context of your specific situation, among other things. It also includes understanding the interrelationships among all the parts of the system, and understanding what actions serve to disrupt the system, and what actions serve to maintain it.

This set of skills enables the work to continue at all, and can cause it to be much more effective and broader in its impact. Many people have built organizations that wither and die when the founder leaves, whether it is a nation, an NGO, an academic department, or a consulting firm. That can be OK, if the founder simply wants to accomplish a single mission, such as getting a single species listed as endangered, or getting a single protected area designated. Many worthwhile and successful groups form, function, and dissolve once their goal is reached. Many worthwhile and successful leaders are leaders only for a single day, or a single issue, or a single project.

But if the leader wants to *keep on* getting species listed, or getting protected areas designated; if the leader has created an ongoing organization with staff, members, beneficiaries, donors, and ongoing projects; if the leader expects the *new* government to continue the programmes developed with the *old* government—then that leader needs to create a system, with an organization and funding and staff or volunteers and community support, to carry on the work.

In addition, many organizations achieve some of their aims but fail to leverage off of the related work of other organizations. Strong systemic skills enable you to achieve much more by capitalizing on the synergies and interrelationships available, to reach your goals.

The leader also must understand the system within which he or she operates, including the political, community, religious, social, economic and historical factors that go into creating the situation as it is now, in order to best move to the situation as it is desired to be. In other words, from the mechanics of how a bill becomes law, to metaphors of organization (are you 'growing a business' or 'fighting the enemy'?), the leader must understand the interrelatedness of these factors.

HOW DO LEADERS LEARN THESE SKILLS?

We believe the best way to learn these skills is by experiencing and practising them in a safe learning environment, apart from the daily activities of one's regular work, with a group of like-minded and equally dedicated people from around the world. Since 1997, we have worked with hundreds of participants in the Smithsonian Environmental Leadership Course, an eleven-day residential workshop offered annually in the US and elsewhere. Based on the outline above, the course comprises 16 modules, beginning with self-knowledge tools such as behavioural profiles and discussions of desirable leadership characteristics and concluding with group presentations incorporating team skills, negotiation, and metaphor.

Along the way, participants learn to define and evaluate the desired outcomes; understand and apply the fundamental presuppositions of leadership; create and maintain personal states of excellence and call upon them at will; elicit detailed information through specific human language patterns (independent of the actual language used, e.g. Spanish or English or Farsi or Sinhalese); notice and 'edit' their own behaviours when needed; create and maintain focus (noticing and managing congruity and incongruities); break information into appropriate 'chunks' for easy transmission and understanding of the intended audience; negotiate and resolve conflicts; develop suitable missions and visions for self and organization; prioritize actions and see from other people's perspectives; create compelling futures, for self and organization; create practical strategies for achieving desired outcomes (modelled on the strategies applied by Walt Disney); apply the various 'logical levels' to solution of a problem (from 'skills' up to and including the spiritual dimension); and understand and create effective verbal, visual, and kinaesthetic metaphors to aid communication and help reach goals easily.

We spend more time learning and practising leadership skills than we do talking 'about' leadership, although we do talk enough to set the scene and provide a context for how we and others have applied these skill sets to environmental issues in the past. In addition, most participants have some personal experience of leadership (positive or negative) to draw upon, and everyone knows something about some important environmental leaders through the world media who can serve as other models of effective behaviour.

We use the example of Dr Karl-Henrik Robért, the Swedish founder of the international environmental management programme called The Natural Step, as a continuing teaching tool during the class, through his audio and video presentations of how he founded that organization in the 1980s (see http://www.naturalstep.org for more information). He is an unique example, we know of an environmental leader, who successfully applies all four skill sets we teach in an ongoing way.

Most of the learning during the workshop is acquired doing actual exercises on each of the specific skills and then applying them to environmental problems or issues. Participants are asked to bring case studies or problems to share with their colleagues and the instructors, and are encouraged to use these real-world situations as topics for the various exercises. That way, participants will not only have learned a new skill, they will have accomplished important tasks towards reaching their goals as well, during the workshop.

We believe especially in an environmental context, that, it is important to keep a positive focus. That is, it is important to focus on what you *want* to happen, not on what you *do not want* to happen. Therefore, we teach explicit tools for developing desired outcomes, measuring their accomplishment, correcting the course when you get off track, maintaining focus and support during the process, and communicating success to supporters, donors, staff, volunteers, and community members.

Any of the topics could be expanded into a week-long workshop by itself. The instructional pace is intense, and for some participants it is quite challenging because the training is in English (although participants are encouraged to use their own languages to conduct the exercises). Beginning in 2006, all of the graduates are supported by an ongoing on-line Leadership Blog, which presents continuing learning activities and insights, and offers a way for graduates to stay in touch and practice their skills as part of a learning community.

WHAT NEXT?

As one of the participants in the 2006 session of the course pointed out, none of these skills is restricted to the practice of environmental leadership—they are skills *any* kind of leader must have. And that is the main point: Many environmental issues are large-scale, long-term, and cross boundary. Their resolution requires collaboration, negotiation, and focus in addition to developing technology and collecting data, and requires leadership in addition to good science. This leadership may come from scientists, and it should be so in many cases. For example, scientists must take the lead in developing consensus data and models of biodiversity, global climates, agricultural programmes, social equity programmes.

But scientists alone cannot solve the earth's problems, whether they be biological, climatological, agricultural, or social. We will need political and social and economic and academic and community leaders too, with finely-tuned leadership skills that foster collaboration, negotiation and focus. We all are proud to have at least a small role in helping this to happen.

REFERENCES

Cherniss, Cary and Goleman, Daniel (2001) *The Emotionally Intelligent Workplace*, Jossey Bass.

Goleman, Daniel (1997) *Emotional Intelligence*. Bantam Press.

Goleman, Daniel (2000) *Working with Emotional Intelligence*. Bantam Press.

Dilts, Robert (1996) *Visionary Leadership Skills*. Meta Publications.

Part V

Environment for Global Well-Being

EDITOR'S INTRODUCTION

Looking at the world trends of lack of compassion and tolerance, increasing ambition for more and more, and increasing terrorism challenged the very survival of human race, the greatest need of the present era is to rise above the national boundaries and look for paths to global well-being. The Indian philosophy of treating the whole world as one family and having a collaborative approach in all spheres of life, is worth to be mentioned here as a model for the western world to follow. Foreseeing the climate induced disasters for the planet earth, now the whole world is trying to join hands together in combating the possible calamities and sharing their resources, and sharing their resources. The slogan of 'survival of the fittest' is now considered as being generated by the basic instincts or tendencies which have to be transformed into the principle of 'togetherness' and collaboration . This is the only way out to save the planet in a sustainable way for global well-being. So the question is that we may be moving towards globalization:

The following two chapters beckon towards global well-being which is possible only if people develop a positive environmental appraisal and rise above the materialism which cannot lead the world towards a green sustainable future.

11

Positive Environmental Appraisal and Spiritual Mindset for Global Well-Being

Beyond Materialism

PARUL RISHI

The natural environment in which one survives and the social, psychological or spiritual environment which provides the sense of contentment, fulfilment and ultimate happiness, are precious to be preserved for enduring happiness and well-being. Looking at the world trends of lack of compassion and tolerance, unbridled ambitions and increasing terrorism, challenging the very survival of human race, the greatest need of the time is to rise above the national boundaries and look for paths to global peace and happiness.

This chapter reviews the cross-continental values for materialistic and non-materialistic possessions, especially between Western and South Asian countries, and further suggests how a judicious move from materialism to supra-materialism associated with positive environmental appraisal (PEA) and Indian spirituality can provide road map to global happiness. It further suggests that complimenting and supplementing precious cultural and spiritual values across countries can develop the world as a whole by earning as well as living potential for ultimate happiness.

The linkage between human needs and environmental appraisal has given an undue right to the living creatures, to assault the environment for planning their pleasures. In spite of the foresightedness of the living creatures, there was an utter failure on their part to assess the cumulative cost of such actions. This cost may be greater for some and the lesser for others in the short-term, but in the long run, all persons share the common world for their survival and sustenance, so any negative effect of their actions on the planet will have global consequences affecting all. Kaplan (1993) stated about more than a decade before that "environmental planning and economic development depend upon environmental appraisal". In simple terms, how one perceives the current environmental issues and develops his cognitive interpretations, determine the way environmental planning will be done and issues of economic development will be taken up. This whole decision-making process varies from country to country in spite of the definite knowledge that action of any one country will affect the planet earth as a whole.

Human possessions are divisible into materialistic and non-materialistic categories. All materialistic possessions are priced, directly or indirectly, however, non-materialistic possessions like natural environment in which one survives, and the social, psychological or spiritual environment which sustains the sense of contentment, fulfilment and ultimate happiness these do not have any specific price, yet they are precious to be preserved for sustainable happiness.

THE BLIND RACE

World is divided in terms of economic development and not by environmental conservation and quality. These worldwide divisions give supremacy to the economic development over environmental preservation, and as a result, developing and underdeveloped countries strive to follow the western model of development. Therefore, GNP, the index of economic growth is very much in the news to indicate country's economy as against the gross environmental conservation index. It is very much linked to consumerism, investments and government spending. Higher GNP is the positive indicator of desirable growth. Opening of shopping malls, personal wealth and purchase of consumer products indicate existing economic situation of a country. Many people from the developing countries visiting western countries, return with a enchanted impression of prosperity and wish that their country should also start running in this blind race of consumerism and commercialization for raising their economic standards.

WHEN ENVIRONMENT PEEPS IN?

In the early 1990s, environmental issues started peeping in the developmental race and the culture of environmental economics also started gaining popularity. The world started viewing environment in terms of loss of trees and wildlife, land degradation, air quality, water purity, waste creation and management etc., despite the great economic cost for their management. The developed world began realizing the social trap, i.e., the temporal cost which the other parts of the world will have to bear as a result of overconsumption of the few fortunate. However, the cause of worry was not overconsumption of few fortunate but the increasing pace of commercialization in the developing nations with huge populations to come in line with the developed nations in the race of economic development. Durning (1991) argues perceptively that overconsumption by the world's fortunate segments is an environmental problem unmatched in severity by anything but perhaps population growth.

TOWARDS SOCIAL DILEMMA!

This whole issue can be best interpreted in terms of theories of social dilemma which is a difficult choice whether to act in public interest or self-interest. In times when resources were unlimited and population was limited, one could afford to choose the both, but in the continuously reversing situation, one has to choose between the two undesirable outcomes. Every country was earlier in the position to exploit the natural resources as very few countries either knew how to exploit them, or having limited material resources to make that happen. Besides, the available population was also limited. In this case, developed countries were particularly capable of using more and more wood for material consumption, more and more automobiles for luxurious driving, more and more air conditioners for combating the natural temperature conditions, and many more consumer goods which may have high consumerist cost in environmental terms to be shared by rest of the world at the middle to lower end on the scale of consumption. It is easier and more rewarding at least in the shortrun, to engage in self-serving behaviour, than to behave in public interest (Gifford, 1987). With the limited resources, public interest behaviour is expensive, difficult, time consuming and less immediately rewarding than self-interest behaviour.

Dawes (1980) has suggested two major characteristics of social dilemma. The first one states that each participant receives more for the self-interest choice than for the public interest choice. However, the second one states that participants' as a group benefit more by public interest choice than by self-interest choice (defect). Frequent defections destroy the resource. The similar situation occurred in the world when few countries were acting in self-interest for resource consumptions, they were at the advantageous end, but when 'defectors' increased, there was a threat to the resource and environment at large. At that stage, the earlier 'defectors' who were getting the major share, became the propagators of preserving the precious environment for the common cause. They started initiating various environmental summits and advised whole of the world to act in the public interest for the sustenance of resource.

However, in the race of more and more consumption at one side and preserving the precious environment on the other, the western countries forgot about the most invaluable element of human life, namely, happiness. They forgot that overall quality of life of all the human beings in this universe is interwoven with and interconnected by getting the larger share of resources, one can not purchase subjective well-being or happiness.

SUBJECTIVE WELL-BEING AND HAPPINESS

In recent years, psychological or subjective well-being (SWB) is the focus of research attention (Diener and Diener, 1995; Diener, 2000). Psychological well-being resides within the experience of the individual (Campbell et al., 1976). It is a person's evaluative reaction to his or her life—either in terms of life satisfaction (cognitive evaluations) or affect (ongoing emotional reaction).

Diener and Diener (1995), however, held that little attention has been paid to whether the predictor of psychological well-being differs in various cultures. They further added that variables that influence people's evaluations of their life do vary across culture. While working in the area of psychological well-being in India the role of personality and social support in that matter has been highlighted (Nathawat, 1988; Nathawat and Rathore, 1996). Whereas, Diener et al. (1989) held that people who are successful at attending frequent positive affect will be happy.

The above views are understandable in the light of the observation, that economic affluence may not necessarily be related with happiness. Campbell et al. (1976), for example, disclosed that during the period 1957–1972, when most of economic and social indicators were moving rapidly upward, the proportion of population of the USA who was described as 'very happy', declined steadily. They further revaluated that 17 per cent of life satisfaction or happiness is predictable from 10 demographic indicators in national probability sample.

HOW TO CHOOSE?

The macro-level choices and decision-making for environmental issues are primarily governed by the pre-existing social, cultural and spiritual mythology too, besides the principles of logic and rationality. In the United Sates, consumerism and success are identified with each other and they have their own logic behind that. Japan's sacred treasures are television, air conditioning and automobiles (Durning, 1991). In a developing countries, situation varies from place to place. In the country like India, where spiritual and cultural ethos were quite strong at one point of time, the socio-psychological effects of overcommercialization and globalization are now becoming visible in major urban centres, though still at a very small scale as majority of population still lives in the rural areas or countrysides, far away from the availability and economically affording capacity of the consumer products. The result may be economically very dissatisfying, but environmentally, quite satisfying. The environmental appraisal of such areas may differ quite substantially from the point of view of local residents and the economic analysts with all rationality and logic. The human development reports also proclaim that the challenge of development is to improve the quality of life.

Now, the question is from which perspective? Quality of life may have the social perspective, the psychological perspective, the environmental perspective or the economic perspective or even the spiritual perspective. Rachel Kaplan (1991) raises a question, "Are billions of people of the developing world without modern consumer products and transportation systems, in poor psychological health?" In this regard, the case of Bhutan is worth to mention here.

The Case of Happy Little Kingdom of Bhutan

One of the world's least developed countries, Bhutan is worried about what globalization may bring about and it is determined to protect its unique culture and social values by entrenching them in such terms as the world may understand and respect in quantifiable measures. By developing measures of progress that account properly for the country's social, cultural, and environmental assets as well as its economic development, the country is following through on the 1972 declaration made by His Majesty the King Jigme Singye Wangchuck: "Gross National Happiness (GNH) is more important than Gross National Product". The four pillars of GNH, as mentioned by Thinley, the Prime Minister, are the promotion of equitable and sustainable socioeconomic development, preservation and promotion of cultural values, conservation of the natural environment, and establishment of good governance. While these ideals are already ensconced in state policy, the Bhutanese government is increasingly interested in measuring and even quantifying its progress according to these values.

While household incomes in Bhutan remain among the world's lowest, life expectancy increased by 19 years from 1984 to 1998, jumping to 66 years. The country, which is preparing to shift to a constitution and an elected government, requires that at least 60 per cent of its land remain forested, welcomes a limited number of wealthy tourists and exports hydropower to India. "We have to think of human well-being in broader terms", said Lyonpo Jigmi Thinley, Bhutan's home minister and ex-prime minister. "Material well-being is only one component. That doesn't ensure that one is at peace with one's environment and in harmony with each other". It is a concept grounded in Buddhist doctrine, and even a decade ago it might have been dismissed by most economists and international policy experts as naïve idealism. Bhutan's effort, in part, is aimed at avoiding the pattern seen in the study at Harvard, in which relative wealth becomes more important than the quality of life.

"The goal of life should not be limited to production, consumption, more production and more consumption", said Thakur S. Powdyel, a senior official in the Bhutanese Ministry of Education. "There is no necessary relationship between the level of possession and the level of well-being." Mr Saul, the Canadian political philosopher, said that Bhutan's shift in language from 'product' to 'happiness' was a profound move in and of itself.

Bhutan needed to ensure that prosperity was shared across society and that it was balanced against preserving cultural traditions, protecting the environment and maintaining a responsive government. Still, Bhutan's model may not work for larger countries and even in Bhutan, not everyone is happy. Members of the country's delegation admitted their experiment was very much a work in progress, and they acknowledged that poverty and alcoholism remained serious problems and the pressures of modernization are also increasing. Bhutan linked itself to the global cultural pipelines of television and the internet in 1999, and there have been increasing reports in its nascent media of violence and disaffection, particularly among young people.

Bhutan's example, while still a work in progress, is serving as a catalyst for far broader discussions of national well-being. Around the world, a growing number of economists, social scientists, corporate leaders and bureaucrats are trying to develop measurements that take into account not just the flow of money but also access to health care, free time with family, conservation of natural resources and other non-economic factors.

WEST RETHINKS SUBJECTIVE HAPPINESS

In the World Values' Survey, a project underway since 1995, Ronald Inglehart, a political scientist at the University of Michigan, found that Latin American countries, for example, registered far more subjective happiness than their economic status would suggest. In contrast, countries that had experienced communist rule were unhappier than non-communist countries with similar household incomes—even long after communism had collapsed. “Some types of societies clearly do a much better job of enhancing their people’s sense of happiness and well-being than other ones even apart from the somewhat obvious fact that it’s better to be rich than to be poor,” Inglehart said. Even more striking, beyond a certain threshold of wealth people appear to redefine happiness, studies suggest, focusing on their relative position in society instead of their material status. One approach is to study how individuals perceive the daily flow of their lives, having them keep diary-like charts reflecting how various activities, from commuting to watching cricket, make them feel.

A research team at Princeton is working with the Bureau of Labour Statistics to incorporate this kind of charting into its new time use survey. “The idea is to start with life as we experience it and then try to understand what helps people feel fulfilled and create conditions that generate that”, said Alan B. Krueger, a Princeton economist working on the survey. Subjecting students to more testing in order to make them more competitive, may equip them to succeed in the American quest for ever more income but that benefit would have to be balanced against the problems that come with the increased stress imposed by additional testing.

Britain is also developing an ‘index of well-being’, taking into account not only income but mental illness, civility, access to parks and crime rates. British officials released their first effort along those lines, a summary of ‘sustainable development indicators’ intended to be a snapshot of social and environmental indicators like crime, traffic, pollution and recycling levels. “What we do in one area of our lives can have an impact on many others, so joined-up thinking and action across central and local government is crucial”, said Elliot Morley, Britain’s environment minister.

Messinger, the Director of Industry, takes the pulse of his country’s economy, sifting streams of data about cash flow to generate the figure called gross domestic product, but for nearly a decade, he has been searching for a better way of measuring the quality of life. "A sound economy is not an end to itself, but should serve a purpose, to improve society" Messinger said. The new well-being index, Messinger said, will never replace the GDP. For one thing, economic activity, affected by weather, labour strikes and other factors, changes far more rapidly than other indicators of happiness. The team of researchers plan to integrate the findings with measurements of education, environmental quality, 'community vitality' and the responsiveness of government. Similar initiatives are also under way in Australia and New Zealand.

In the United States there have been a few experiments, like the Princeton plan to add a happiness component to labour surveys. But the focus remains on economics. The Census Bureau, for instance, still concentrates on collecting information about people's financial circumstances and possessions, not their perceptions or feelings, said Kurt J. Bauman, a demographer there. But there was growing interest in moving away from simply tracking indicators of poverty, for example, to looking more comprehensively at social conditions.

That shift was evident at the conference on Bhutan, organized Colman, who is from Nova Scotia. Participants focused on an array of approaches to the happiness puzzle, from practical to radical. John de Graaf, a Seattle filmmaker and campaigner trying to cut the amount of time people devote to work, wore a T-shirt that said, "Medieval peasants worked less than you do". In an open discussion, Marc van Bogaert from Belgium described his path to happiness: "I want to live in a world without money". Al Chaddock, a painter from Nova Scotia, immediately offered a suggestion: "Become an artist". Other attendees insisted that old-fashioned capitalism could persist even with a shift to goals broader than just making money. Some experts, who attended the week-long conference, questioned whether national well-being could really be defined. Just the act of trying to quantify happiness could threaten it, said Frank Bracho, a Venezuelan economist and former ambassador to India. After all, he said, "The most important things in life are not prone to measurement—like love". "Even in a very miserable condition you can be very happy if you are grateful for small mercies," said Siddiqur Osmani, a professor of applied economics from the University of Ulster in

Ireland. "If someone is starving and hungry and given two scraps of food a day, he can be very happy."

World Happiness Survey

Recent research on indicators of happiness across countries has further proved that economically underdeveloped countries are much ahead on indicators of happiness in spite of being underprivileged may be owing to PEA which is beyond materialism. Bangladesh is the happiest nation in the world. The United States, on the other hand, is a sad story: it ranks only 46th in the World Happiness Survey. That way the US is behind India, the fifth happiest place in the world, and others including Ghana and Latvia, Croatia and Estonia. Research under the auspices of London School of Economics tries to look into the link between personal spending power and the perceived quality of life has conclusively shown that money can buy everything save happiness.

The study revealed that people in Bangladesh, one of the poorest countries in the world, derive far more happiness from their small incomes than, for example, the British (32nd on the list) do from their relatively large bank balances. In fact, people in most rich countries including Austria, The Netherlands, Switzerland, Canada, Japan and others are much unhappier than their poorer counterparts in countries like the Dominican Republic and Armenia. Most unfortunate, however, are Russians and the people in some other parts of the former Soviet Union. They are neither rich nor happy, indicates the World Happiness Survey. Slovenia, Lithuania, Slovakia, Russia, Ukraine, Belarus, Bulgaria and Moldova follow the United States in the list to bring up the rear. The study shows that although the British have twice as much money to spend in real terms compared with 40 years ago, their perceived quality of life has not improved.

Earlier surveys revealed that many Britons thought money could bring happiness. The new study shows that such a link still exists in poor countries because a small increase in income can mean large improvements in lifestyle. However, beyond a certain income level that direct relationship breaks down. According to the research, happiness in rich countries now is far more dependent on close personal relationships, good health and job satisfaction.

People in Britain are generally less happy than they were ten years ago. Two thirds would rather see the environment improved than have more economic growth and personal spending money. said Robert Worcester, visiting professor of government at the LSE and a co-author of the study.

The researchers have concluded that although Britons are rich compared with most other countries, many suffer from an emotional poverty caused by consumerism and the breakdown of family life. We are being seduced by an economic juggernaut and our personal needs are not being met, said Nic Marks, a social sciences researcher at Surrey University who also worked on the report (*New York Times*, 4 October 2005).

This research has lead the whole of the West to rethink on core culture of developing countries to find out the hidden force which governs positive environmental appraisal and subjective well-being or happiness. In this regard, the case of *Dayalbagh* a spiritual centre in India is also worth to be sited here.

REFERENCES

Bhaskaran, K. (1991) "Meditation from a Mental Health Prospective". *Indian Journal of Psychiatry*, 33, 87–94.

Campbell, A., Converse, S.E., and Rodgers, W.L. (1976) *The Quality of American Life*, New York, Russell Sage Foundation.

Dawes, R.M. (1980) "Social Dilemmas", *Annual Review of Psychology*, 31, 169–193.

DEI (2005) *Progress of Education in Dayalbagh*, Agra: Dayalbagh Educational Institute Publication.

Diener, E., Sandvik, E., and Pavot, W. (1989) "Happiness in the Frequency, Not Intensity of Positive Versus Negative Affect". In F. Sttrack, M. Argyle, and N. Schwarz (eds.) *The Social Psychology of Subjective Well-being.*

Diener, E., and Diener, M. (1995) "Cross Cultural Correlates of Life Satisfaction and Self Esteem". *Journal of Personality and Social Psychology*, 68, 653–663.

Durning, A. (1991) "Asking How Much is Enough". In L.R. Brown (ed.), *State of the World 1991*), pp. 153–169, New York, Norton.

Formen, J. (1974) *Transcendental Meditation*. New York, Duttoh.

Fromm, E. (1976) *To Have Or To Be*, Suffol Abacus.

Ghosh, A.I. (2005) "Doncept of alienation and Better Worldliness". In proceedings of National Symposium on 'Education for Better Worldliness', October 1–2, 2005, p. 26.

Gifford, R. (1987) *Environmental Psychology: Principles and Practice*, Massachusetts: Allyn and Bacon, Inc.

Kaplan, R. (1991) "Environmental Description and Prediction: A Conceptual Analysis". In T. Garling and G. Evans (eds.), *Environment, Cognition and Action: An Integrated Approach*, pp. 19–34. New York, Oxford University Press.

Kaplan, R. (1993) "Environmental Appraisal, Human Needs, and a Sustainable Future". Garling, T. and Golledge, R.G. (eds.), In *Behaviour and Environment: Psychological and Geographical Approaches*. Amsterdam, Elsevier Science Publishers B.V.

Maslow, A.H. (1943) "A Theory of Human Motivation", *Psychological Review*, 50: 370–396.

Mathur, S.N. (2005) "Education for Better Worldliness". In proceedings of National Symposium on 'Education for Better Worldliness', October 1–2, 2005, p. 28.

Mehra, G.P. (2005) "Transforming India into a Major Global Education Centre for Better Worldliness". In proceedings of National Symposium on 'Education for Better Worldliness', October 1–2, 2005, p. 39.

Nathawat, S.S., and Rathore, S. (1996). "Influence of Hardiness and Social Support on Well-being in Elderly Men and Women". *Indian Journal of Clinical Psychology*, 23, 92–101.

Nathawat, S.S. (1975) "Yoga and Behaviour Therapy". In A.K. Sinha (ed.), *Perspectives in Yoga*, pp. 83–93, Varanasi, Bharat Manisha.

Nathawat, S.S. (1988) "Factors Influencing Social Well-being in Research Students in India". Abstracts of XXIV International Congress of Psychology (Abstract No. F 40), Sydney, Australia.

Nigam, A. (2005) "Integrated Personality Development for Better Worldliness". In proceedings of National Symposium on 'Education for Better Worldliness', October 1–2, 2005, p. 27.

Parikh, J. (1991) *Managing Yourself: Management by Detached Involvement*, Oxford, Blackwell Publishers.

RSS (2002) *Dayalbagh: A Brief History and Development*, Agra: Radhasoami Satsang Sabha Publication, 14th Edition.

RSS (2005) *Discourses on Education in Dayalbagh*. Agra: Radhasoami Satsang Sabha Publication.

Satsangi, P.S. (2003) "Reality and Truth, vis-à-vis Wisdom". In Proceedings of International Conference of the Association of Asia Pacific Operational Society, New Delhi, December 8, 2003.

Swarup, A. (1932) *Interpretation of Preaching of Bhagwat Gita,* Agra: Radhasoami Satsang Sabha Publication.

Toffler, A. (1970) *Future Shock,* New York, Random House.

Verhelst, T.G. (1987) *No Life Without Roots: Culture and Development,* (Trans. B. Cumming, 1990) London, Zed Books.

World Commission on Environment and Development (1987) *Our Common Future.*

Websites Referred

Andrew C. REvkin (2005) *A New Measure of Well-Being from a Happy Little Kingdom,* New York Times, Published: October 4, 2005. http://www.nytimes.com/2005/10/04science/04happ.html

Jacqueline Weaver (2005) "Meditation Associated with Increased Grey Matter in the Brain" http://www.eurekalert.org/pub_releases/2005-11/yu-maw111005.php

Proceedings of the Second International Conference on Gross National Happiness at Nova Scotia from June 20th–24th 2005. http://www.gpiatlantic.org/conference/papers.html

12

We Care for the Planet! Linking Spirituality with Worldliness at Dayalbagh

PARUL RISHI

If one wants to look at the environmentally healthy and sustainable, economically self-reliant, politically untouched, socially and psychologically satisfying and spiritually escalating temple of wisdom, worldliness and spiritual vision, one has to consider the case of Dayalbagh, which literally means 'Garden of the Merciful'. Located within the important urban centre, Agra, the city of Taj, this small place is in existence since 1915 as the self-contained colony of the followers of Radhasoami sect. Those who believe in the sect hold that there is one true Supreme Being, who is the eternal source of all spirituality and the creator of the entire universe, and that the spirit of human beings is an emanation from the same Supreme Being. It also emphasizes that humans should make it the aim of their life to develop all their faculties namely, physical, mental and spiritual, throw off all lethargy and develop *work is worship* as the mission of life. To procreate this mission, the 'Garden of the Merciful' was tastefully created and crafted to provide the people honest simple living, employment through small-scale industries, high level educational facilities up to university level, health facilities and residential complex.

The fifth leader of this faith Sir Anand Swarup Kt. known as revered *Sahabji Maharaj* stated before the Lindsay Commission in 1931: "Our idea is that the rise and fall of nations takes place as a result of the presence or absence of extraordinarily gifted souls and such souls take birth only in communities and countries where they get adequate facilities for carrying on their work. That is why we in Dayalbagh are making an attempt to maintain pure spiritual atmosphere typical of India and simultaneously to provide freedom and facilities to work as are available in the West".

Dayalbagh has been described as a village as far as peace and quietness is concerned and an ideal city as far as providence of basic amenities and cleanliness is concerned. Students, scientists and devotees keep themselves engaged in their respective activities away from the turmoil of the world. Neither does wealth flow here nor people live with any deprivation; nor are there any big palaces and mansions, nor any dilapidated huts either; neither is anyone great or big nor anyone small and insignificant, and if anyone is honoured here more than others, it is he who works better than others. Dayalbagh belongs to every resident, while no resident has any kind of property here.

On hearing this and looking at the surroundings and people, Nehru (1956) commented that "countries advance not an account of their size but by the manner its people lead their lives and by the character they possess and by the skill of their hands and intelligence they possess to do things. If India would make progress, it would do so because such people live here and not because thirty six billion people live here". Simple and economical lifestyle with limited use of consumer luxuries is the motto here so that physical, mental and spiritual faculties can be developed in harmony. People are encouraged to participate in physical exercise followed by social service in agricultural fields to remain physically healthy and productive. "Fatherhood of God and brotherhood of mankind" is the basic principle of this faith which teaches universal brotherhood pervading all prejudices and superstitions regarding caste, creed, nationality, or colour. No wasteful expenditure is suggested for birth, death or marriage. Looking at the trends of modern civilization, one finds that economic welfare of the society and the individuals comprising it, and the organization of a socialistic pattern should be the predominant objects of all social activity. But if one tries to find out whether economically better off individuals and communities are happier than those that are not so well placed, we would discover that they are not necessarily so. Happiness, where material well-being is concerned, is mostly a question of values and the world would be spared of many a quarrel and consequent misery, if only a sense of right values is developed. The teachings of the sect have emphasized the idea that the most important part of man is his spirit (*atman*) as distinguished from his body and mind, and therefore, the cultivation of spiritual values and giving them their due place in the economy of human life would alone lead to harmony between man and man, between man and universe which alone can make the world a better place to live in.

At Dayalbagh, year in and year out, the day begins with silent meditation followed by the singing praises of the Merciful Lord. The members begin their work in the pure natural environment of agricultural fields and then spend the day in honest and diligent discharge of their duties in the offices or workshops, educational institutions, hospitals, etc., and the day closes with humble thanks giving at the Feet of the Gracious Lord by the residents, men and women, old and young. Every member tries to work for the common and the leader of the community sets an example of selflessness and devotion to duty.

THE LESSONS TO LEARN

The case of Dayalbagh clarifies the relation between economic and environmental indicators and human condition. The balance of all is *better worldliness*. At one side, there is the sacred treasure of Japan—'The Automobile' or the well developed motorized fast transportation system of other developed countries, at another side is the choice of Dayalbagh, the temple of wisdom, to allow limited motorized transportation within the sacred site in spite of having the residence of about five thousand people and thus avoiding the possible air and noise pollution. Bicycle is the major mode of transport. Another sacred treasure of Japan—the television, is also not visible there, although people are in good touch with the whole of the world by weekly common display of useful programmes, thus avoiding its evils and preserving its information values. The most interesting feature is the good affording capacity of many of the people living over there and it is their willing choice to enjoy the limited worldliness within a pious behavioural setting. A limited enjoyment of consumer products, to the extent that person may not became the slave of it, is the beauty of that socio-behavioural setting which promotes PEA.

In the times of automation, social service in terms of physical work in agricultural fields (the major source of self-reliance of foods products) is visibly reflected in health and longevity of people living over there. The whole world is now realizing the need for using bicycle to conserve oil and avoid air and noise pollution, besides keeping one in good health. The whole world is now realizing the ills and evils of television and its possible socio-psychological and health effects on the generation which is going to take the charge of the world in the time to come. The whole world is now realizing the need for social support network and strong family bonds to obtain solace, sympathy and support for emotional release which counters the psychological distress arising out of complexities of life, rising temptations and unrealistic desires. The whole world is now realizing the need for spiritual wisdom to obtain contentment and satisfaction with the present and move towards psychological well-being and total quality of life from different perspectives. It is possible only if world can proceed for materialism to supra materialism and ultimately achieve the balance between the two through PEA.

MATERIALISM VS SUPRAMATERIALISM

Materialism appears quite attractive from the distance but to get its real taste, one has to work a lot in a limited time. It generates emotional pressure and anxiety and makes coping a little more difficult which gives rise to irritation and further inability to cope. It is said that stress is indispensable for meeting the ever increasing challenges of life. It starts which 'Eustress', the positive stress which is required in promoting the performance. However, when it crosses the limit of 'Eustress' and associated peak performance stage and enters the state of distress and associated burn out state, it becomes very difficult to sense. Once a person gets the taste of achievement and associated sensory and mental pleasure, he just runs after that to get more and more. The outcome of this race for more is faced by no one but the person himself in terms of psychosomatic symptoms like cardiac problems, hypertension, tension headache, diseases of gastro-intestinal system etc.

Materialism requires money, technology and people to use the acquired material. However, use of materials to serve the necessity is not materialism in the real sense of the term. It is the complete dependence of people and a sort of psychic attachment with the luxury associated with it along with the desire to use it as a symbol to indicate the social status.

Materialism can provide only temporary happiness and satisfaction, and there is something beyond materialism. Everyone agrees that there was the time of simplicity, with fewer consumer products and things, fewer technologies and a slower pace of life. This simplicity now seems to be a blessing in disguise as it has some invisible virtues. Verhelst (1987) describes them as "vitality, a taste for life, human qualities and a sense of the sacred that leave a lasting impression of hope". He further says that harsh social struggle do not get in the way of "a sense of celebration, human tenderness and ultimately, the desire for social harmony". The case of Dayalbagh also embodies the simplicity model for ultimate satisfaction.

THE FAILURE OF SUCCESS

While the world has progressed economically, technologically or scientifically, human beings seem to have regressed despite their intellectual and materialistic progress. As an outcome of the revolution of progress and consequent mass production and consumerist system, human beings are getting more and more mechanized, specialized and standardized. Mass media dominated by commercial interests is challenging the strings of prevailing socio-economic system (Parikh, 1991). Absence of value-based education and the enlightened family culture further promotes hardiness and alienation among people. Jagdeesh Parikh has described this situation as the 'failure of success'. Unprecedented techno-economic and scientific achievements are trapping people in acquisitive culture intended for sensory gratification of lower level and selfish attitudes. A high standard of living is equated not with better 'thoughts' but with better consumer products and their possessions. The great tragedy of the civilization is that it has improved the living standard of human beings but not the human beings themselves. The value systems are rapidly declining and human life has already become so invaluable that ordinary deaths go unreported and catastrophes involving even million people could just be considered as a matter of chance. It gives rise to a few very pertinent questions like, Are we heading towards a virtue filled world full of joy compassion, love or concern? Or a just task oriented world? (Nathawat, 1975). There is no doubt that a new world is unfolding its different and never-ending layers of change, the full magnitude of which is impossible for us to comprehend (Toffler, 1970). It seems that most of us are in a state of suffering, a little more or a little less, depending on our circumstances. Only the free-being is exempt from this suffering. In this regard, the following lines are worth quoting:

> There was a Time, when the world was fine
> The birds did chirp, that world was mine
> Now the days that came, aren't any fun.
> We are lost in work, unfeeling, hard and cold
> This is the modern world
> I am told!
>
> —Anonymous

Money, power and prestige can be the symbols of success but not necessarily the symbols of peace and happiness. A popular saying states that they can provide food but not appetite, clothes but not beauty, house but not happy home, books but not wisdom, medicines but not health, beds but not sleep, partners but not love, and above all, the luxuries of life but not happiness in life. Therefore, being in close touch with environment and appraising it positively is the path for global well-being.

SUPRAMATERIALISM IN INDIA

The research conducted in both developed and the developing world proves that there is something beyond materialism and that may be called as *supramaterialism* where, in spite of enjoying the world by pleasures, luxuries and consumables, one is mentally not attached to them. They are being used only as the modes of serving basic requirements as and when required. People are not their slaves rather they are the slaves of human beings to be used as and when required. In spite of the viability of resources to get them, they are being used sparingly and are not used as the status symbols.

Figure 12.1 A Model of Materialism vs. Supramaterialism as Related to Well-being

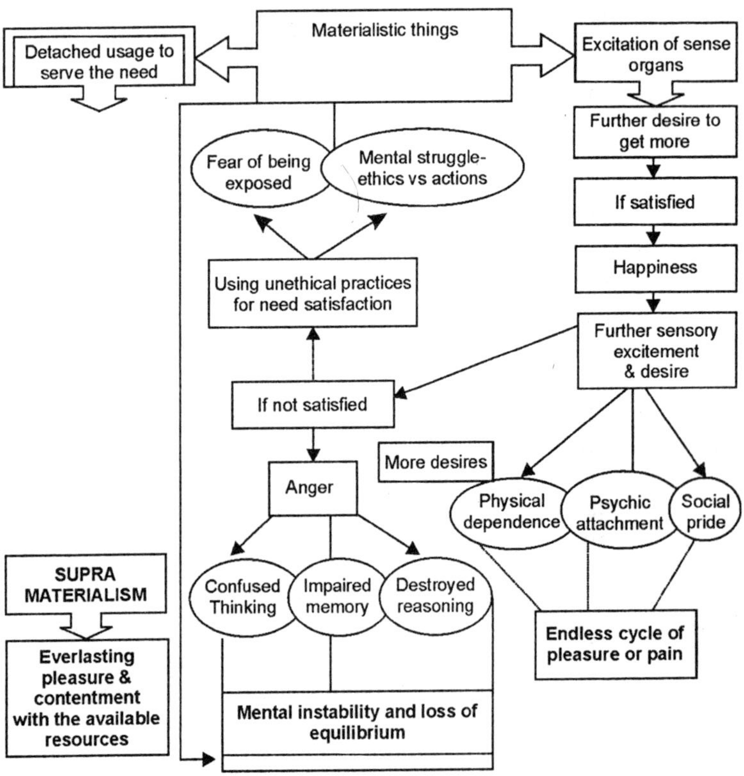

Bhagwat Gita also narrates the importance of supramaterialism for eternal happiness. It states that a person should always concentrate on action and not on the fruits of that action to remain focused on his inner self without disturbing his equilibrium through success or failure and pleasure or pain. This state of equilibrium also known as *Yoga* or the state of restraint from worldly desires and control over mind and body. Yoga is basically a skill involved in performing an action. This is the stage of supramaterialism in which in spite of performing all worldly action using materialistic things, a person is above them as far as attachment to them is concerned. Control over the senses stabilizes the mind. Otherwise materialism excites the sense organs and generate attachment which further creates the desire to indulging in them. Where this desire remains unsatisfied, anger results which confuse the thinking, impairs the memory and destroys the logic or reassuring power, resulting in the loss of mental stability and equilibrium (Figure 12.1).

In this way, materialism and associated desires is an endless process and a root cause of all troubles, whether one obtains the materialistic satisfaction or otherwise is immaterial. Even if it is obtained, it causes physical dependence, psychic attachment and social pride associated with it which may not be permanent and ultimately, will give rise to suffering. So, people with the stable minds are not affected by attachments or pleasures or anger. Neither love overpowers them nor hate. Their minds are away from worldly enjoyments in spite of the fact that they are using them.

SENSORY RESTRAINT VS ACTION

The *Bhagwad Gita* further classifies that sensory restraint and detachment from materialism should not be confused with being actionless. The attributes of nature compel everyone to perform actions or duties. Sitting actionlessly will take the person more towards sensory pleasures, so action or duty with a spirit of detachment or mental restraint, known as *Karma Yoga* is the desired condition of all happiness.

On the other hand, desire and anger are insatiable and are the biggest enemies which are all consuming and polluting that enwrap the world and overpower wisdom. The strength of wisdom flows from the lowest level, i.e., from sense organs to mind, a little higher level to intellect, a little more higher level to spirit, the most powerful superior and mightier of all which can subjugate the mind with its might and crush the desire with its supreme power and make possible the communion with God the Almighty.

This hierarchy is in line with Maslow's (1957) hierarchy of needs. Upto the level of self-esteem are the lower order needs which are associated with desire and sensory pleasure. The whole world strives to move up in the hierarchy of need satisfaction, but upto the level of self-esteem only. The higher order needs of self-actualization which are also known as meta needs are being fulfilled by less than 2 per cent of the population. Once a person is able to overpower the lower order needs, he looks for contentment of higher order needs. There is another hierarchy within the higher order needs too that moves up from mind to spirit through wisdom and intellect and moves up to the ultimate reality of communion with the supreme being through mediation.

In this regard, the step to move up in the hierarchy starts with restraining sense organs and controlling them. Once this is made possible, controlling the mind and intellect comes next and then comes controlling the spirit and making it possible to achieve the ultimate reality or communion with the Supreme Being, the highest form of self actualization. The criticism of Maslow regarding vagueness of this term and achievement of this state even without the satisfaction of lower order needs disregarding hierarchy can be best explained by *supramaterialism*. Self-actualization is basically a state of 'supramotivation' when person is above the level of lower order need satisfaction and need satisfaction up to self-esteem is immaterial for him. If Supramaterialism is viewed from this point of view, self-actualization is independent of the lower order need hierarchy and a construct in itself which its own hierarchy. Any person, who is able to practice action with detachment, taking control of senses, mind and intellect, can achieve this state no matter his lower order needs are satisfied or not. This is the reason why one can find numerous such examples in India where in spite of economic undergrowth and limited resources, people are able to realise self-actualization which is the ultimate reality and the root of enduring happiness.

TOWARDS BETTER WORLDLINESS

Ancient era was the time of better worldliness, a better way of life, where seats of education were the religious institutions and great seers led the world. With development, authority was transferred to kings and the powerful. The result was ego, greed and selfishness in place of contentment. Industrial revolution further increased a cut-throat competition, jealousy and mistrust which snatched the positive human relation between person and person or nation and nation. The need of the time was betterment of world order with peace, contentment and well-being, for which we need great seers imbued with an attitude of selfless service to humanity. One of the greatest Indian seers Sir Anand Swarupji once remarked in 1935. "Religion is the greatest gift of God to humanity. It is the most ancient friend of man and in the coming future it will prove itself to be the saviour of human race". From religion, he never meant the dogma that separates person from the person or creates a mass of superstitions to make the human being a slave and retard human progress. Rather, he meant the spiritual light to purify and illuminate the human heart, to elevate and expand the human soul and to make it possible for human mind to realise the Fatherhood of God and Brotherhood of human beings.

The case of a poor villager or forest dweller, much away from the light of civilization is worth to mention here. If one asks who consoles him, who encourages him and who keeps him moral and honest, the answer will not be philosophies or penal codes but the elementary teachings of religion to sustain him and keep sustained. Education can expand the knowledge and make one logical but not always happy, inventions can make one powerful but not self-controlled and consumer possessions can make one more ambitious but not contented. As worldly possessions are limited, humanity will have to learn the lesson of contentment to achieve lasting pleasure and happiness.

It is now being observed that love of wealth and other earthly possessions, overproduction and cut-throat competition, jealousy and distrust, the products of the passion of greed, are the real cause of all our troubles. For that, there is a need to overcome the lower appetites and satisfy the higher cravings by self-actualization and spiritual insight. Abstract intelligibility, the ideal of science, can never provide permanent happiness to life.

Another great seer of Radhasoami sect mentioned in 1944 about the need for a better world order through PEA. The natural, social and cultural environment around the world is getting degenerated. The world is now witnessing terrible destruction and bloodshed in the form of terrorism at different parts of world. The live examples of mass destruction at the World Trade Centre in 2001 and frequent terrorist' attacks in different parts of India remind us the need for 'a better world order'. From the corner of every country at this time, there is but one cry that such a new order should be established in which people might pass their time in comfort, happiness and peace, and hatred among people and countries may vanish for ever. For establishing this new world order, there is a need for community which is united in purpose, thought and action. There is also a need to conserve the human resources, monitory resources and materials resources with foresight, discipline, self-sacrifice and hard sustained work, applying the principle of economy in implementing any programme or activity.

VALUE-BASED EDUCATION: AN ANSWER TO MANY UNANSWERED ISSUES

The best possible higher education along with a safeguard against purely materialistic approach is the need for a new world order through PEA. This is possible only through value based education to bring about physical, intellectual and ethical integration of an individual with a view to evolving a complete person possessing basic values of humanism, secularism and democracy to counter the social and environmental challenges. More so, it is required to inculcate the spirit of truthfulness, temperance and courage by cultivating the spirit of humanity, simple living, selfless service and sacrifice. This can be realised through promotion of scientific temper along with the background of humanities to retain the human touch, to mellow down the harshness of a mechanical world, and to develop pride in national culture and heritage.

Besides, multireligious learning to develop an attitude of tolerance and a sense of national unity, disregarding the discrimination of caste and class is also to be preached. Such value-based education can generate a consciousness of democratic values to respect the rights of others and an awareness of one's duties and obligations to the society for building a strong character and attaining high ethical standards. It can be seen in line with the latest concerns on the stream of business ethics and behaviour.

MEDITATION: A KEY TO ULTIMATE HAPPINESS

Many people believe that meditation enables a person to reach a state of profound rest, as claimed by popular self-help books (Foremen, 1974). Bhaskaran (1991), has observed that meditation would appear to have preventive potential, though its relaxing effects in stress-induced psychological disorders and in 'Noogenic Neuroses' characterized by dissatisfaction with living in spite of having all the material satisfaction, a sense of meaninglessness in living, a sense of alienation from oneself and one's environment. Meditation may promote the 'Being' mode of living postulated by Fromm (1976), against the 'Having' mode. Meditation would also seem to have the potential for enhancing psychological well-being, especially if we include the spiritual dimension in our concept of happiness. Many Indian sects have promoted meditation as a mode of relaxation, contentment and above all developing communion or oneness with the Supreme Being and Merciful God. Radhasoami sect advises meditation of the holy sound *Radhasoami*. Similarly, Buddhism, Jainism and specifically yogic exercises also have the base of meditation in different forms by contemplating. Meditation improves concentration, purifies the mind, and gives inner peace and strength to face the worldly affairs through detached involvement, thereby promoting better worldliness and consequent happiness.

RECENT LINKAGES BETWEEN MEDITATION AND BRAIN

A study in Harvard Medical School looked at brain parts involved in memory and attention in subjects who meditated. The study found that meditation may aid in brain functions by increasing the thickness of specific parts of brain responsible for memory, vigilance, attention, reasoning and other associated mental processes. University of Wisconsin also conducted studies on Buddhist monks and found similar results to prove the efficacy of meditation in brain functioning (*USA Today*, November 15th, 2005).

Another new study by researchers from Yale, Harvard, Massachusetts General Hospital, and the Massachusetts Institute of Technology shows meditation is also associated with increased cortical thickness. The structural changes were found in areas of the brain that are important for sensory, cognitive and emotional processing. The study further suggested that meditation practice can change anyone's grey matter. The study participants were people with jobs and families. They just meditated on average 40 minutes each day, Magnetic Resonance Imaging (MRI) showed that regular practice of meditation is associated with increased thickness in a subset of cortical regions related to sensory, auditory, visual and internal perception, such as heart beat or breathing. The researchers also found that regular meditation practice may slow age-related thinning of the frontal cortex.

Most of the regions identified in this study were found in the right hemisphere, essential for sustaining attention, which is a central practice of insight meditation. They said other forms of yoga and meditation likely to have a similar impact on cortical structure, although each tradition would be expected to have a slightly different pattern of cortical thickening based on the specific mental exercises involved (*Neuro Report*, November 12, 2005).

MEDITATION FOR SPIRITUAL DEVELOPMENT

The revered leader of Radhasoami sect has mentioned that the corpus of language derived from physical senses pertains to this world, the material plane in which we live, and not beyond that. All modern science also speaks about informatics and systems science. On the other hand, there is another corpus of language derived from esoteric experimentations and research of the saints, particularly in the oriental world, which pertains to regions beyond physical world, by making use of hidden or latent senses which exists in human head and which can be rendered kinetic through meditation and yogic practices uniting the spirit current with the sound current of higher planes (Satsangi, 2003). Spiritual development through meditation is the path for being in touch with the ultimate divine reality and acquiring wisdom, through which highest subjective happiness can be realised.

A LINE OF BALANCE: LINKING MATERIALISM TO SUPRAMATERIALISM

The world should now move from materialism and supramaterialism for achieving the better worldliness and subjective well-being. A careful equilibrium between the two is the source of contentment. Sheer materialism of the West is not very desirable for ultimate happiness as it is promoting many environmental or socio-psychological ills like global warming, environmental degradation and natural resource depletion on one hand, and possessiveness, dependence, and psychic attachment on the other, which may not provide peace and contentment with the life. A number of studies in developed countries with high GNP and material resources and continuously reducing levels of subjective well-being have justified it well.

Total restraint from materialism, i.e., from all materialistic possessions like Buddhist monks and many others may also be practically difficult for people with limited religious inclination and prove to be detrimental to growth and development of a society. The higher indices of Bangladesh and few other countries can not be considered very encouraging until and unless they are linked with the optimal developmental indices too.

This situation may not be true for the developed countries as they have already achieved the higher stages of development. Maslow (1957) also mentions that satisfied need seizes to motivate further. However, development is still a need for Third World to come in line with their privileged partners in the developed world. With that reason, restraining from materialism cannot be stated as the only source of happiness and well-being in the Third Word, rather one should move from materialism to supramaterialism as the most desirable condition, described as *Better Worldliness with PEA.* The key to better worldliness have been described as discipline, self-sacrifice, and hard sustained work with an economical usage of available material resources and unity of purpose, thought and action. It is leading a high quality worldly life of contentment and happiness with humanitarian values and compassion, close family bonds and continuously engaging in selfless action.

CONCLUSION

Looking at the world trends of lack of compassion and tolerance, increasing ambitions for more and more, at one side and increasing terrorism, challenging the very survival of human race on the other, the greatest need of the time is to rise above the national boundaries and work together for positive environmental appraisal, global peace and happiness. Happiness is a subjective feeling of well-being which cannot come from outside. Individuals are described as 'human beings', and as Pareek (1991) says, they should justify the 'being' by rising above the 'doing' and 'having' trap of today's materialistic world.

Indian culture preaches the ideal of *Vasudhaiva Kutumbakam* which means, treat the whole earth is like your extended family and show your compassion to every living being, may be the animals and plants too besides human beings. Extension of the spirit of this statement across countries can promote better world order world wide for global happiness.

Synthesizing the East and the West of the world by linking the earning potential and living potential may be desirable. The West should learn compassion, contentment and selfless service, while third world should learn the spirit of hard sustained work for development. Complementing and supplementing the precious social, cultural and environmental values across countries will develop the world as a whole for progress and happiness. Development with environmental conservation and contentment should be target for which the world should work together for preserving the precious human race and promoting global happiness. This may be the globalization in real sense of the term.

Conclusions

Humanizing the Planet for Global Well-Being

The plant earth needs a human touch. Only then it will ensure the global well-being. For ensuring the well-being of the planet, we have to move from the jungle of concrete to the jungle of green trees; the jungle of materialism to the alter of spiritualism, the struggle for consumables to the peaceful collaboration in sharing resources. There is a need to reverse the growing trends of migration towards urban settlements. It is possible only when there is more and more genuine developmental efforts on the part of government and private sector for raising the earning and living potential of rural communities and establishment of basic infrastructure. It can not be ensured by just flowing national and international money. It requires human beings with a highly dedicated service-oriented mindset, who wish to take initiative in development of rural areas. Development, in this context, does not mean creating the concrete jungle in villages and increasing the availability of consumer goods which are not energy efficient. It means adequate and efficient utilization of locally available resources to create livelihood options as well as developing energy efficient infrastructure which is not detrimental to global climate.

The statements made by India regarding their low per capita emission of greenhouse gases (GHG) appears to be quite appealing. However, every one knows that it is at the cost of poor developments conditions in rural areas which are in abundance over here. Owing to limited electrification, limited modes of transportation utilizing fossil fuel, and limited temptations of rural people due to low economically affording capacity, green-house emissions are low in those areas. India may be considering it as blessing in disguise, however, it is in the same line with what developed world used to think for the developing countries by advising them to reduce their GHG emissions at the cost of their development, so that global climate can be preserved. It is a real dilemma, still to be looked into without compromising with the concerns of the underprivileged as well as the global environment.

To work out this issue, there is a need for more global environmental leaders like Dr Rajendra Pachuri of India, who can look at the problem from humane as well as the technical and social point of view. A culmination of humane, social and technical vision is needed which is to be guided by spiritual wisdom to face any crisis with strength and courage. Besides, just leaders can not do anything without the wholehearted support and collaboration of human. A positive, environment oriented mindset action even at the cost of sacrificing same of the luxuries of life is desirable at the global level for which mass efforts of psychologists around the globe are needed. Many of us can be very positive towards environment, but concrete action is the need of the time to save our planet and ensure global well-being. A major change in lifestyle and habits will be needed to ensure the above.

It is true that individual level efforts add up to give big results, but ethical responsibility of industry in conserving the planet can not be ignored. Industry is the main consumer of fossil fuel resulting in high GHG emissions. Their adoption of energy efficient technology in product design and production process can also create a mass impact on the environment. Transportation system itself is the major source of GHG emissions and industry around the globe has already started taking initiatives to make it energy efficient. However, the issue of concern is the developing world is not able to afford the higher cost of energy efficiency. On the other hand, in a country like India, there is a major need to chalk out a plan for removal of energy inefficient old vehicles from the road. Currently, they just move to the poorer buyers but continue to impact the environment by their high emissions. The pollution check of these vehicles is basically a mockery of the system.

The above analysis finally converges towards humanizing the environment for global well-being. It is said that transportation and communication has reduced the global barriers and the whole world is moving towards oneness.

Beyond this statement the people of the planet earth gather on the same platform surpassing all the national, cultural, social, religious and ethnic boundaries. Now people have started thinking beyond their national concerns to global concerns because they know that human life is interwoven in a similar way as the global climate is. Every single irresponsible action on the part of each person affects the globe, so we all share the great responsibility in conserving the planet for global well-being. All the world is a stage as Shakespeare says, and human beings are the major actors on the stage of global environment. It is their responsibility to act in a way which ensures the sustainable future. This whole scenario reminds the Indian thought *'Vasudhaiva Kutumbakam'*, which means, the whole earth is like an extended family. Every one will have to collaborate and take care of each other when needed, share the available resources, celebrate the pleasures and share the pains. This is the humanized approach to global environment for well-being and sustainable future.

Index